When Benjamin Franklin Met the
Reverend Whitefield

Witness to History

Peter Charles Hoffer and Williamjames Hull Hoffer, *Series Editors*

ALSO IN THE SERIES:

When
BENJAMIN FRANKLIN
Met the
REVEREND WHITEFIELD

Enlightenment, Revival, and the Power of the Printed Word

PETER CHARLES HOFFER

The Johns Hopkins University Press | *Baltimore*

The Johns Hopkins University Press
2715 North Charles Street
Baltimore, Maryland 21218-4363
www.press.jhu.edu

Library of Congress Cataloging-in-Publication Data
Hoffer, Peter Charles, 1944–
    When Benjamin Franklin met the Reverend Whitefield : enlightenment,
revival, and the power of the printed word / Peter Charles Hoffer.
      p.   cm. — (Witness to history)
    Includes bibliographical references and index.
    ISBN-13: 978-1-4214-0311-3 (hardcover : alk. paper)
    ISBN-10: 1-4214-0311-0 (hardcover : alk. paper)
    ISBN-13: 978-1-4214-0312-0 (pbk. : alk. paper)
    ISBN-10: 1-4214-0312-9 (pbk. : alk. paper)
    1. Franklin, Benjamin, 1706–1790. 2. Franklin, Benjamin,
1706–1790—Philosophy. 3. Whitefield, George, 1714–1770. 4. Whitefield,
George, 1714–1770—Philosophy. 5. United States—Intellectual life—18th
century. 6. Enlightenment—United States. 7. Great Awakening. 8. United
States—Church history—To 1775. 9. Scientists—United States—Biography.
10. Evangelists—United States—Biography. I. Title.
    E302.6.F8H67 2011
    973.3092'2—dc22   2011008670

A catalog record for this book is available from the British Library.

Special discounts are available for bulk purchases of this book.
For more information, please contact Special Sales at 410-516-6936 or
specialsales@press.jhu.edu.

The Johns Hopkins University Press uses environmentally friendly
book materials, including recycled text paper that is composed of at
least 30 percent post-consumer waste, whenever possible.

CONTENTS

When Benjamin Franklin Met the
Reverend Whitefield

Prologue

A Momentous Meeting

ONE RAINY MORNING in early November 1739, two men met in the parlor of a rented house on Second Street in Philadelphia. One was a twenty-four-year-old Anglican missionary, George Whitefield, late of Bristol in the English West Country. His reputation for dramatic preaching had preceded him, and he had already pitched his message of rebirth in Christ to thousands from the gallery of the Philadelphia Court House. There he cried—and his listeners cried with him—at the dire prospects for the unconverted and the assurance of grace that surely would come to the convert. Tall and stooped, with one lazy eye giving his oval face a cross-eyed appearance, he now gazed on the other man with compassion belying his years.

The other man in the room was the thirty-three-year-old Philadelphia printer and entrepreneur Benjamin Franklin. Franklin's solidly built figure was well known in the city, and he lived only a few blocks away from the house that Whitefield rented. Curious to see and hear the preacher's performance, Franklin stood in the rain one evening and listened. He was amazed at the way Whitefield's words carried in the open air and the effect they had on his audience. Franklin decided to meet Whitefield in private.

The room was cluttered with donations in kind—clothing and furnishings—for Bethesda, Whitefield's orphanage in Savannah, Georgia, to which he soon would return, but not before he spent another two weeks in Philadelphia converting sinners. That morning, the two men agreed that they could be of use to one another. Whitefield would provide copies of his journals and sermons, and Franklin would publish them. In the partnership struck that day—a partnership of published words between equals—the greatest revival preacher of his day and the man who would come to symbolize Enlightenment science and rationality began one of the most unique, mutually profitable, and influential friendships in early American history.

It is somewhat out of fashion, at least academic fashion, to speak of "representative men," though in the early years of historical writing, books of "parallel lives"—the comparison of leaders from various countries—were the model for all biography. In the nineteenth century, a great man theory of history dominated the thinking of many scholars. As Ralph Waldo Emerson wrote in "The Uses of Great Men" (1855), "Nature seems to exist for the excellent. The world is upheld by the veracity of good men: they make the earth wholesome. . . . We call our children and our lands by their names." In Emerson's America, biographies of great men were best sellers.[1]

Biography is alive and well today, but the great man theory of history so popular in the nineteenth century ill fits modern history teachers' emphasis on the plain folk. Without meaning to sound old-fashioned, this volume rests on the assumption that there are people who both represent their times and alter them in crucial ways. Franklin and Whitefield were two such men, even though they seemed polar opposites in their thinking. Franklin was worldly and philosophical, and he believed that human endeavor could improve standards of living everywhere. The hidden springs of nature Franklin found through experimental science. Whitefield was passionate, spiritual, and convinced that man was powerless to save himself. The hidden world of the spirit Whitefield knew through faith. Franklin had founded highly secular institutions: a library, a philosophical society, a hospital, and a college. Whitefield was one of the founders of English Methodism, a reform movement within the Church of England that looked to the Bible for everyday guidance and called for a return to early Christian values. It is almost too easy to find such differences in the two men's thinking and outlook.

But Franklin and Whitefield had much in common, enough so that they saw themselves in each other. Both rose in station in life from the sons of, respectively, a candle maker and a tavern keeper to consort with lords and ladies. Both kept detailed diaries, a measure of their close monitoring of their public images. Both were performers, the public arena their stage. While it may be too simplistic to say that they loved publicity, they certainly understood its uses.

By focusing our story on Franklin and Whitefield, we can see into the intellectual heart of their world, the Anglo-American empire of the eighteenth century. Not so long ago, English and North American historians celebrated the Anglo-American transatlantic connection. The turn-of-the-century "imperial school" of Atlantic historians insisted, "to England alone, of all the civilized powers that bordered on the Atlantic in the seventeenth century, do we trace our descent as a nation." More modern studies of the Atlantic World have a broader focus, and rightly so, for the shipping lanes of the Atlantic brought together African and Native American peoples as well as transplanted Europeans. Imperial fortunes, in both senses of the word, depended on the slave societies of the Caribbean more than on more northern and more European settlements.[2]

Franklin and Whitefield were truly Atlantic World figures. They were known, were admired, and made an indelible impression on the British colonies and the home country, linking the English-speaking peoples on both sides of the ocean together. In different ways, they elevated the culture of that world. Over the course of their careers, they would cross and recross the Atlantic nearly a dozen times and, through great effort, intelligence, and ambition, reinvent themselves into the foremost spokesman for technological progress and the premier revivalist preacher, respectively. Individually and together they represented what is now a lost world—the Anglo-American empire—at its height.

In 1739, the Anglo-American Atlantic World was a busy, prosperous place. Its American provinces had made the home country wealthy. In turn, the provinces' population and standard of living improved, Pennsylvania itself drawing immigrants from England, Wales, Scotland, Ireland, various parts of Germany, France, and the Iberian Peninsula. Not all of the immigrants thrived. Slaves and indentured servants faced a far bleaker future than newcomers with resources. But the lure of riches and the push of want drove hundreds of thousands of Europeans to brave the Atlantic crossing and the uncertainties of life in a new land—as many as 315,000 people between 1700 and 1770.[3]

Human traffic along the Atlantic highway moved in both directions, though the volume westward was far greater than eastward. Still, Anglo-Americans returned to their British homes (or those of their ancestors), visited the metropolis, or sought audiences with those in power in the home country. By the first decades of the eighteenth century, the Atlantic had ceased to be a barrier and had become a well-traveled highway of people and ideas. The remarkable commercial and military success of the British Empire and its cultural coherence depended on the traffic along this highway. Indeed, the British people attained "a collective identity" in the period before the American Revolution, an identity based in part on the "social openness" and "intellectual and scientific achievement" the Atlantic highway fostered.[4]

The British and American sides of the ocean were hardly cultural equals. The British North American colonies were "a borderland, a part of the expanding periphery of Britain's core culture." But in the middle of the eighteenth century, the "provinces," including the North American colonies, had entered something of a "golden age" of cultural expression. Print tied the two sides of the empire together. Reading literacy on both sides of the Anglo-American Atlantic had taken a huge jump forward at the beginning of the eighteenth century. "From 1730 to 1750 . . . the quantity of reading possibilities was growing"—and growing rapidly. Many Americans could read (though fewer could write). "Not only did more people read, people read more."[5]

Rising literacy and the proliferation of print media went hand in hand. The printing press became an almost spiritual force, turning spoken words into imperishable lines on the page. Newspapers had come into vogue. Magazines of opinion were common. Book production and marketing were a growth industry, much like e-books today. Books published in England were commonplace in American libraries. Somewhat less so, books published in America found their way to British shops and shelves. Words in books and newspapers and magazines transmitted ideas, "particular meanings . . . adopted by the speech community and imposed in turn on its members." English-speaking colonists "sent their children back to England to be educated" to master words. England was the mecca for the provincial would-be gentleman or intellectual. From England and Scotland American schools and churches

recruited learned men, and they, along with the native-born colonists, produced a rich treasury of literature, scholarship, and political essay.[6]

For all our infatuation with the e-world, texting, blogging, e-book publishing, and otherwise hurling our words into the electronic ether, the *words
themselves* are the sources of our fascination. Franklin and Whitefield understood the power of printed words. Franklin would publish Whitefield's journals and sermons and cover the preacher's travels in the *Pennsylvania Gazette*.
Through Franklin's later efforts, a subscription raised enough money to build
Whitefield a meetinghouse when his manner of preaching and his views on
salvation caused most of the city's clergy to deny him their pulpits. Though
the friendship would wax and wane (Franklin was not persuaded that he
needed to be born again in Christ, but Whitefield's manner was always gentle
and affecting and Franklin could not stay mad at him), their usefulness to
one another never flagged. Franklin became Whitefield's promoter and publicist in America, and Whitefield's peregrinations made Franklin's newspaper
must reading for everyone curious about the Great Awakening of religiosity.
Franklin turned Whitefield's aural power into print. The selling of Whitefield
helped make Franklin a wealthy man—wealthy enough to surrender everyday business affairs to others and set about his scientific experiments and his
sponsorship of a multitude of urban schemes.[7]

Franklin was entrepreneurial and inventive when it came to marketing
words. He gained appointment as colonial postmaster and used it to send his
Pennsylvania Gazette all over the colonies. He sought and obtained the post
of printer to the government of the colony. From the October 1729 issues, his
first after driving its former publisher into bankruptcy, he sprinkled crime,
sex, and humor throughout the issues of the *Gazette*—all surefire popular topics. He sought and published "scoops" and even "stole" stories from his rivals.
Using pseudonyms for letters he wrote himself, his correspondence section
wryly and pointedly nailed his competitor's foibles. So, too, his *Poor Richard's
Almanack*, from 1732 to 1757, supposedly compiled by one Richard Saunders,
made fun of all manner of subjects, including other printers. Much of its content was borrowed (even the name Richard Saunders was that of an earlier
English almanac compiler). The sayings were lifted from other publications,
spiced by Franklin's own turn of phrase. Still, the yearly publication became
a best seller. Almost everything that Franklin wrote found a ready audience.[8]

Whitefield was also aware of the importance of the print media. He needed it to extend the reach of his preaching. Though he spoke extemporaneously, he prepared text assiduously. He then supplied it to the newspapers. "Press coverage helped insure Whitefield and his revival widespread popular acceptance." Though he seemed above the crass consumerism that seemed to have taken hold of England in the first half of the eighteenth century, Whitefield had made himself (even more than the content of his sermons) into a consumer product. Without the aid of print, Whitefield's ministry might have been stillborn. Even words that set thousands to swooning were written on water without press coverage and print publication. William Seward, a confidant, convert, and former stock issue salesman, helped Whitefield turn oratory into print in the same way that over two decades earlier Seward had sold South Sea Company stock (the period's version of junk bonds) to the unwary. "Notices" of Whitefield's appearances became "advertisements" for the man and the message. With Whitefield's initial hesitation to commercialize his revival overcome, Seward acted the role of press agent.[9]

Whitefield's words carried across the ocean. From England to America news of Whitefield's open-air ministry traveled in press releases. In America, editors eager for newspaper copy and increased subscriptions saw the opportunity to carry news of the Whitefield phenomenon. When Whitefield himself followed, generally to carry on the Methodist missionary work of John and Charles Wesley in South Carolina and Georgia and specifically to raise funds for an orphanage in Savannah, Georgia, the press coverage was transatlantic. News of his triumphal tour filled the London newspapers and magazines. He then decided that he must carry the mission to all the British North American colonies, and did, uniting them, and bringing out crowds, in an unprecedented fashion. Indeed, in the eighteenth century the only event comparable to his tours in uniting the disparate colonies and stirring the passions of the mass of people was the American Revolutionary crisis.

Franklin was a transatlantic traveler himself, and he understood how the printer and publisher's skills could draw the home country and its colonies closer. When not yet twenty, he spent a year in England mastering the typesetter's trade. Returned to Philadelphia the next year, he found sponsors, then partners, then friends to build a business around words. Words spanned the ocean. When he started the *General Magazine* (one of his few unprofitable ventures), he offered its contents and took as its subject "all the British plantations in America." For Franklin, the wider the scope of his publishing ambitions, the better they suited his view of himself as a transatlantic figure.

Commerce held the project together, for shopkeeper Franklin marketed more than newspapers, but even this commerce in clothing, foodstuffs, and consumer durables was dependent on transatlantic suppliers and markets. And how better to ensure that buyers knew the latest imports on one's shelves than to advertise in the *Gazette*?[10]

In sum, both Franklin and Whitefield had learned that the Anglo-American Atlantic World could be more closely knit together by words. They were part of the process of course, but they could not have had the success they enjoyed without other supportive innovations in information technology. The introduction of regular packet boat service, the rise of newspapers, and the investment in port facilities enhanced the speed and spread of news. The coming of war heightened all of these, as merchants needed to know where danger lurked on the ship routes, at the same time as it made all communication and personal travel more perilous. And war is what came in the summer of 1739.[11]

This is not a book about Franklin the revolutionary diplomat or Whitefield the established leader of the evangelical movement, though they would fill these roles later in their lives. It is not the first volume of a full-fledged dual biography, though one could surely be written and would be most welcome. It is primarily a book crafted to return its readers to a time and place in the colonial period when revolution was the furthest thing from Franklin's mind and Whitefield's revivalism was still fresh. It is about a time when the Anglo-American empire was full of possibilities and opportunities for those with ambition and vision.

The sources for this study abound in Franklin's and Whitefield's own works, including Franklin's autobiography and collected papers and Whitefield's journals and sermons, as well as the burgeoning newspaper culture of the period and the documentary records of the metropolitan center and its colonial periphery.

Some technical matters require our attention in these sources. Old-style English official dating, in which the new year began on March 25, I have changed to conform to modern-style dating (the year beginning January 1). English authorities changed from old style (OS, the Julian calendar) to new style (NS, the Gregorian calendar) in 1752. A hybrid, in which dates from January 1 to March 25 were rendered with a slash (e.g., February 1, 1730/31), was in use during the eighteenth century. I have also silently modernized difficult-to-decipher eighteenth-century contractions, awkward grammar, and incorrect spelling in the primary sources.

The two men cut a wide swath through their times and correspondingly in the secondary (scholarly) literature. There is a cottage industry in Franklin biographies and collections (over 350 to date), to which the three-hundredth anniversary of his birth added excellent titles. Whitefield has not fared so well in the realm of biography these days, perhaps because his intense piety does not quite lend itself to modern scholars' ears as it did to those in the early nineteenth-century Second Great Awakening (when a number of major biographies of him appeared). Religious writers still find his rhetoric worthy of study, but more attractive to modern biographers are Whitefield's theatrical and commercial sides.

Finally, a word or two about the structure of the book. My story is both chronological and topical, hence after a chapter devoted to the title event—the meeting of the two men and the arrangement between them—the book's next four chapters are divided between narrative and analysis. Chapter Two brings both men to their meeting place, combining biography and narrative history. The next two chapters explore in detail some of the central themes of Whitefield's and Franklin's early writings, respectively. Chapter Five returns to narrative, placing the two men in the central intellectual controversies of their times: Whitefield in the Great Awakening and Franklin in the Scientific Revolution. That chapter concludes with the last years of the two men's collaboration and the most important of all controversies in the anglophone world, the protests against Parliament. An epilogue muses on the theme of modernity and assesses Franklin's and Whitefield's contributions to our own times.

A Partnership of
one Mutual Convenience

THE SPRING OF 1739 came late to Philadelphia. When it did come, the rain never seemed to end. The city once again buzzed with rumors of war between England and Spain. Conflict between the two European superpowers dated back to the sixteenth century and came to a violent climax in the defeat of the Spanish Armada (1588) by the English fleet and the weather in the English Channel. Never again the equal of the English at sea, Spain still drew great wealth from its Caribbean, Mexican, Central American, and South American provinces, and the English continued to prey on Spanish treasure ships, slave ships, and settlements. Peace was never really achieved in American waters between the two nations. The episodic and inconclusive Anglo-Spanish war of 1722–1729 interrupted efforts to achieve some quiet, if not amity, between the countries. Disputes over who would and who could carry slaves from Africa to the Spanish colonies in America (the English claimed the right to do so, but in 1739, the Spanish closed the lucrative trade to English ships) fomented further conflict. The war would come in July, when George II allowed his navy to resume offensive operations against the Spanish.

News of the impending conflict was a staple at the taverns and inns on

Market (formerly High) Street in Philadelphia, on Dock Street in Boston, and in Bristol and London, England. No one could tell what effect the war would have on trade, but in British ports ship captains and owners began outfitting their sloops and barks for privateering raids on Spanish merchant shipping. War could be a bonanza or a catastrophe, depending on the success of British naval efforts, the luck of the privateers, and a list of imponderables including disease, weather, and the prices for imports and exports in distant markets.[1]

The fall brought to Philadelphia relief from the pestilent and muggy summer, as well as news of victories over the Spanish, toasts to the King, and the arrival of a celebrity from England—George Whitefield. He had already caused a stir in South Carolina and Georgia the year before, preaching in the open air to hundreds, crowds including slaves and their masters, men and women, Europeans and Native Americans. Like Fortinbras in Shakespeare's *Hamlet*, Whitefield's return to the colonies was much awaited.

Some American observers were skeptical of all the reports, including those of gatherings of thousands to hear Whitefield in the fields around Bristol and London. Boston newspapers were especially quizzical. In mid-September, the *Boston Evening-Post* published a series of "Queries to Mr. Whitefield" copied from a London newspaper: "It would require much more room than we have to spare to give a concise account of the movement of the Methodists during this month. Preaching in church-yards, commons, large spaces, and even in the streets of the city of London, are such new things, and things so little reconcilable to any method formerly in use amongst us, that it is no wonder that it alarmed those who were not in the secret, or that it procured the following letter to be addressed to a certain gentleman who is held to be the author of this new sect." Whitefield was hardly the "author" of Methodism—that plum would go to his mentors, John and Charles Wesley—and the Methodists did not formally depart the Church of England until 1791, after John Wesley's passing. But the Puritans in Boston, themselves something of a sect, were worried. It was a tradition within New England Puritan churches to address pointed questions to ministers before they were hired. The newspaper querist put the old-style interrogation to new use: "Queries to Mr. Whitefield . . . be pleased to specify, I. What are the principles, doctrines, articles of faith, motives and etc. which this extraordinary Light reveals, after what manner they come into the mind, and by what mark or character you distinguish them from the delusions of fancy or worse temptations."[2]

George Whitefield in London in about 1739. No longer a "boy wonder," he would become a leader of the Great Awakening. Portrait by John Faber Jr., published by James Hutton, after G. Beard mezzotint.

Benjamin Franklin's Philadelphia newspaper was more welcoming. His *Gazette* reprinted the account of Whitefield's revival meetings that Seward published in the London *Daily Advertiser*. Two years earlier, Franklin reprinted a London piece on the young prodigy preaching in the city: "a young gentleman distinguished in his piety, very eminent in his profession." On May 3, 1739, Franklin reported that the boy wonder had become a man with a following. "The Reverend Mr. Whitefield preached to about 10,000 people, at Kennington Common. This day he is to preach at Wapping Chapel for the benefit of the Orphan House in George, and at Kennington Common again in the evening." Whitefield was ostensibly collecting funds for the orphanage as well as harvesting converts. "In the evening he preached at Kennington Common to about 20,000 people, among whom were nearly forty coaches, besides chaises, and about one hundred on horseback, and tho' there was so great a multitude, an awful silence was kept during the whole time of singing, prayers, and sermon." One can only imagine the throng; Franklin's readers surely did. Nothing in anyone's experience compared to it. The sheer numbers were staggering (even if Franklin was credulous in reporting what Seward related to the London newspapers). The only comparable gatherings were London street mobs, and even these never quite matched the throng at Whitefield's appearances.[3]

In October, Franklin reported that Whitefield was on his way to Philadelphia before returning to Georgia, he "choosing (as he therein alleges) to go thither by the way of Philadelphia." Though the *Gazette* regularly reported news from England, taking the London papers off the packet ships as they arrived and freely reprinting items (no permissions sought or given), Franklin was paying especially close attention to Whitefield's travels. He must already have had an inkling that publishing Whitefield would be good for both men.[4]

Whitefield arrived in the vicinity of the city on the last day of October and entered it on the second of November, accompanied by Seward, who would ensure that the press coverage was adequate and properly shaped. A small entourage of converts completed the traveling party. Whitefield was not yet married (he would be, quietly, in 1741), and his family life remained something of a mystery in his time and after. Like a whirlwind, however, he swept doubters, sinners, and skeptics up in his train.[5]

Whitefield had a keen eye for places and spaces. Though still a young man, only a few weeks away from his twenty-fifth birthday, he had already traveled across an ocean three times, seen war and peace, and in the process had learned to select those sites best suited to his needs. Whitefield's Church of England colleague Archibald Cummings opened the doors of Christ Church to the young visitor, but within a week its pews were so filled to overflowing with the curious and the converted that a more commodious site was needed. Whitefield selected the balcony of the colony's Court House as his new, outdoor pulpit. Standing at the corner of Second and Market Streets in the busiest part of the city, the Court House was one of the most imposing buildings in Philadelphia. Constructed in 1709 of brick, steeply gabled, its four stories topped by a bell tower, it was the highest structure in the city until the completion of the bell spire of Christ Church in 1753. At the west end of the courthouse was a gallery to both sides of which a flight of steps led. The steps would later become the notorious scene of fisticuffs on election

Friends Meeting House *(left)* and the Philadelphia Court House *(right)*. On the landing at the top of the courthouse steps Whitefield preached to thousands. Historical Society of Pennsylvania.

days. Around the gallery, on the broad street, was a great space, part of which was a viewing area for the stocks and a whipping post. At six in the evening of November 9, 1739, a cross section of the city's population had gathered there, summoned by the bell in the courthouse. Used to notify the populace of danger, this time the alarm was not one of fire or invasion but a warning to the unconverted and the sinful that the pit of hell yawned open at their feet.[6]

Whitefield's journal, both a throwback to the first Puritans' desire to register in minute detail their spiritual state (the only way to gain some "assurance" that they had been among the chosen few saved by God's mercy) and a commercial project (for Whitefield already knew the importance of publicity in his ministry), recorded the incidents of his week-and-a-half stay in the city. As he later wrote, "every particular dispensation of divine providence has some particular end to answer in those to whom it is sent." He threw himself into the arduous tasks of public preaching three times a day, private meetings with other members of the clergy, audiences with individuals and families, singing, praying, baptizing, and catechizing, hoping that "my heart may be made meet" for such an ordeal. The journal also chronicled his travels throughout the city, for "going abroad, if duly improved, cannot but enlarge our ideas, and give us exalted thoughts of the greatness and goodness of God." But the most important passages narrated his mission. "Our Lord was with us as we came on our Way; our hearts burned within us whilst we talked to one another, in psalms, and hymns, and spiritual songs." As recorded in the journals, his travel and his travail always had a biblical parallel, what historians of American religion have called a "typology." He was the second coming of Moses. "Oh how gloriously must the children of Israel pass through the wilderness, when they saw God's presence go with them." And like a latter-day Moses, Whitefield led his flock through that wilderness toward a promised land.[7]

The news of Whitefield's triumphs whetted the curiosity of his fellow ministers, and each day he "met with some gracious souls, who discoursed with me sweetly concerning the things which belong to the kingdom of heaven." Whitefield took care of more mundane matters as well. For himself and his followers he "hired a house at a very cheap rate." Franklin reported Whitefield's choice: "in second street (the same in which Capt. Blair lately dwelt)," almost adjoining the Court House. To it came throngs of the curious and the penitent.[8]

Evening open-air revival meetings marked the high point of Whitefield's visit. He stood on "the court house stairs" a few steps from his rental and could be heard all the way to the wharves. On Thursday, November 8, he "read prayers and preached, rather to a more numerous congregation than I have seen yet." With his accustomed modesty he estimated that "about 6000 people" gathered in the streets below the courthouse gallery. "The inhabitants were very solicitous for my preaching in another place beside the church." As in Bristol and London, where Whitefield preached to multitudes in fields and on hillsides, "the generality of people" of Philadelphia seemed to prefer to hear the word under nature's ceiling rather than in church. The next night, the throng grown to eight thousand, he again ascended the courthouse steps to preach. "I never observed so profound a silence before my coming. All was exceeding hushed and quiet. The night was clear, but not cold. Lights were in most of the windows all around us, for a considerable distance. The people did not seem weary of standing, nor was I weary of speaking."[9]

Whitefield rubbed some of the ministerial fellowship the wrong way, but that was nothing new. He confided to his journal (hence to its thousands of future readers) that the "generality" of the Church of England "do not preach or live the truth as it is in Jesus," that "Papists" were mislead by their priests, that the Quakers should show by outward signs what their "inner light" revealed to them, and that "in bearing my testimony against the unchristian principles and practices of the generality of our clergy" he had to speak with "zeal." He was already engaged in controversies with Anglican authorities in Bristol and the bishop of London over his ministry, the former threatening to excommunicate him if he continued to preach without a license from the Church of England (Whitefield was ordained but did not have a pulpit assigned him in the city), and the latter reminding ministers not to show so much "enthusiasm" in their ministry (a term as used then meaning raving radicalism, thus a clear slap at Whitefield). The next year would find him engaged in a running fight with some in the Methodist movement over doctrine (though he professed to love the Wesleys, the founders of the movement, as brothers) and with various conservative Congregationalists and Presbyterians over preparation for conversion. But this mission was a success as far as he was concerned. The following year he recalled that when "I stood upon a Balcony on society Hill, from whence I preached . . . and felt somewhat of that Divine Fire," many were "converted to my ministry" and thousands seemed "laboring under deep convictions" how "the Lord Jesus made himself manifest to their souls."[10]

Whitefield's repertoire of dramatic gestures and intonations was not quite matched by his repertory of sermons. As Franklin later wrote of them in his *Autobiography*, a practiced listener could tell which of the sermons were new and which were repeated. The latter had a polish that the former lacked, and Whitefield delivered the old ones with greater panache. He practiced these, but gave them extemporaneously, the tears and other dramatic touches real.[11]

The sermons were memorable experiences for the auditory. New England revival leader Jonathan Edwards looked forward to the arrival of Whitefield in Massachusetts, writing to the Englishman, "It has been with refreshment of the soul that I have heard of one raised up in the Church of England to revive the mysterious, spiritual, despised, and exploded doctrines of the gospel, and full of a spirit of zeal for the promotion of real vital piety. . . . I hope this is the dawning of a day of god's mighty power and glorious grace to the world of mankind." After hearing Whitefield preach, Edwards's wife, Sara, recalled, "It was wonderful to see what a spell he cast over an audience by proclaiming the simplest truths of the Bible. I have seen upwards of a thousand people hang on his words with breathless silence, broken only by an occasional half-suppressed sob. . . . A prejudiced person, I know, might say that this is all theatrical artifice and display, but not so will anyone think who has seen and known him."[12]

Whitefield's style seemed an effusion of the spirit. As Josiah Smith, witness to his preaching in South Carolina, recalled, "With what a flow of words, what a ready profusion of language, did he speak to us upon the great concerns of our souls." Nathan Cole, a Connecticut farmer who traveled for miles to hear Whitefield speak, reported that the minister's words and gesture combined in an almost "angelick" *mise-en-scène*, "and my hearing him preach gave me a heart wound."[13]

Even some of those "prejudiced people" Sara Edwards decried conceded Whitefield's oratorical powers. The English lexicographer and wit Samuel Johnson quipped that Whitefield "would be followed by crowds were he to wear a night-cap in the pulpit or were he to preach from a tree." After hearing one of Whitefield's performances, the Scottish skeptic philosopher David Hume joked, "Stop, Gabriel, stop, ere you enter the sacred portals and yet carry with you the news of one sinner converted to God." New York City Anglican minister Richard Carlton was not so amused by Whitefield's message to the city's bondmen, in large measure because he feared that Whitefield's words would whip the lowest rung of the city's people to fever pitch. "Not that I should think Mr Whitefield to be so extreamly wicked as to promote the destruction of this city," but slaves who had thoughts of rebellion might have been urged on by Whitefield's offhand comments, such as, "I have wondered that [mistreated slaves] . . . have not more frequently rose in arms against their owners."[14]

Drawn to the early November 1739 gatherings by curiosity as much as anything else, Franklin found himself spellbound. He did not come to be converted. He was never much of a believer. As he wrote for his own benefit in 1728, "But since there is in all Men something like a natural Principle which inclines them to Devotion or the Worship of some unseen Power; And since Men are endued with Reason superior to all other Animals that we are in our World acquainted with; Therefore I think it seems required of me, and my Duty, as a Man, to pay Divine Regards to Something." Not exactly an endorsement of Christian doctrines, this. There were many supreme beings, but of that being who watched over Franklin, he conceived, "for many Reasons that he is a good Being, and as I should be happy to have so wise, good and powerful a Being my Friend, let me consider in what Manner I shall make myself most acceptable to him. Next to the Praise due, to his Wisdom, I believe he is pleased and delights in the Happiness of those he has created; and since without Virtue Man can have no Happiness in this World, I firmly believe he delights to see me Virtuous, because he is pleas'd when he sees me Happy." Whitefield's God was pleased when he was worshiped with perfect faith. Franklin's God was happy when Franklin was happy.[15]

Franklin was not easily fooled by someone who claimed a direct line to God's ear. He had a sharp eye for dissimulation and would not brook it. "I believe it is impossible for a man, though he has all the cunning of a devil, to live and die and villain, and yet conceal it so well as to carry the name of any honest fellow to the grave with him." If no one else could see behind the facade, Franklin thought he could. And he liked what he saw and heard of Whitefield.[16]

Much later in his life, he recalled with near-perfect clarity Whitefield's eloquence: "He preached one evening from the top of the court house steps, which are in the middle of the market street, and on the west side of second street which crosses it at right angles. Both streets were filled with his hearers to a considerable distance." Franklin's curiosity then took a characteristically scientific turn.

> Being among the hindmost in Market Street, I had the curiosity to learn how far he could be heard, by retiring backwards down the street towards the [Delaware] River, and I found his voice distinct till I came near Front Street, when some noise in that street, obscured it. Imagining then a semicircle, of which my distance would be the radius, and that if it were filled with auditors, to each of whom I allowed two square feet, I compute that he might well be heard by more than thirty thousand. This reconciled me to the newspaper accounts of his having preached to 25000 people in the [London area] field.[17]

That was also a lot of potential readers. Franklin knew a good deal when he heard one, and he approached Whitefield with an offer. Franklin's rival, Philadelphia printer Andrew Bradford, had already published two of the sermons soon after Whitefield arrived. Franklin knew he had to act quickly, lest Bradford win the right to publish the journals in the colonies as well.[18]

Whitefield's journal never mentioned his meeting with Franklin or the precise terms of the agreement the two made. I surmise that the two met privately before Franklin heard the minister speak to the multitude. Franklin knew where Whitefield lodged. Franklin would not have joined the crowds singing and praying. He would have waited for a quiet moment. When he entered, the two men would have sized one another up. Franklin was a little under six feet tall, solidly built and already inclining somewhat to the stocky. With his "thin tight lips and his high domed forehead," he was an imposing figure. Perhaps to the meeting he wore a replacement for the "coat lined with silk" and the "fine holland shirt" stolen from his house earlier that year. Franklin was no longer the poor apprentice looking for work. In face-to-face encounters he strove never to give offense, a manner that outsiders might think "complaisant," but was in fact a mask.

Whitefield was round-faced, stooped, and thin, with a long sharp nose and bulbous lips. His most striking feature was a lazy left eye that drifted in toward his nose, giving him a cross-eyed look. Later in life, he too would grow heavyset. Franklin's portraits were kind to him, the prosperous businessman giving way to the wise and prudent statesman. Though Whitefield would commission a series of portraits, they were never flattering, and the caricatures of him as a religious fanatic were far more popular.[19]

The preacher then asked the well-to-do printer for a contribution to the orphanage. Franklin looked about at the objects that charity had deposited with Whitefield for transport to Savannah—candlesticks, dishes, nails, shot, lead bars, powder, blankets, bed stuffing, cloth and clothing of all kinds, curtain fabric, and even silk. The catalogue he later reported implies that Franklin tarried a while, his and Whitefield's attention focused on the baubles of this world rather than Franklin's fate in the next. Whitefield inquired after Franklin's religious beliefs, and Franklin probably dodged the question. Whitefield saved souls, but Franklin thought his own already safe. Now the business portion of the meeting commenced, and it concluded fruitfully. A week later the *Gazette* revealed that "the Rev. Mr. Whitefield having given me copies of his journals and sermons, with leave to print the same; I propose to publish them with all expedition, if I find sufficient encouragement." At this time, most book publication relied on "subscriptions," that is, advance orders. "Those therefore who are inclined to encourage this work, are desired speedily to send in their names to me, that I may take measures accordingly."[20]

Franklin knew that Whitefield's ministry was controversial, and how that controversiality might affect sales. Franklin had been involved in one religious imbroglio already. Early in the 1730s, Franklin attended five consecutive

Sunday services at the Presbyterian church. He found the minister Jebediah Andrews uninspiring, however, and thereafter spent his Sunday mornings elsewhere. Late in 1734, Samuel Hemphill arrived to assist Andrews, and Franklin returned to hear the new minister preach. Hemphill's sermons, delivered "with a good voice," were "most excellent discourses." Hemphill had little use for dogma and less for predestination or the mysteries of religion. He preached the practical virtues. This bothered some of the elders, and they called for his ouster.

Franklin, under an assumed name, published vigorous defenses of Hemphill, and these included an assault on the clergy. Franklin's April *Defense* sold well, and his July *Some Observations* was a best seller. "For soon after, Hemphill was represented by several Ministers to be a New-Light Man, a Deist, one who preach'd nothing but Morality, a Missionary sent from Ireland to corrupt the Faith once delivered to the Saints; in short, he was every thing a persecuting Spirit could invent." For those who accused Hemphill, to Franklin's ear at least,

> if they gave any Credit to the Sermons upon which they afterwards condemned him; and which they were pleas'd to declare they believ'd to be genuine, and read to 'em as they were preached, they had then the highest Reason to object to the Credibility and Faithfulness of the Evidence; seeing the Sermons plainly prov'd most of the Evidences to be false. . . . And here I am sorry, that I am obliged to say, that they have no Pattern for their Proceedings, but that hellish Tribunal the Inquisition, who rake up all the vile Evidences, and extort all the Confessions they can from the wretched Object of their Rage, and without allowing him any Means of invalidating the Evidence, or convincing 'em of their own Mistakes, they assemble together in secret, and proceed to Judgment.[21]

Franklin's defense of Hemphill lay not on religious grounds (that is, that Hemphill's theology squared with mainstream Presbyterianism) but on the grounds that Hemphill's critics had violated the canon of basic fairness (i.e., that they had been immoral in their tactics). Franklin was also spurred by the stance that Andrew Bradford, his newspaper competitor and nemesis, had taken. Bradford published the ecclesiastical trial, defending the presbytery. The controversy ended with a whimper when Hemphill's enemies asserted that he had borrowed a little too much of a sermon from a published source. Exhausted by the tempest, his supporters melted away, and the young minister left the colony. Franklin moved on to other less controversial causes, until he decided to champion Whitefield.

The arrangement—Franklin to publish, Whitefield to supply clean and edited copy—would hold for the next thirty-one years. For Franklin, though he never accepted Whitefield's invitation to confess sin and be born again, genuinely respected the minister. Indeed, as Franklin later boasted,

> The multitudes of all sects and denominations that attended his sermons were enormous, and it was matter of speculation to me, who was one of the number, to observe the extraordinary influence of his oratory on his hearers, and how much they admir'd and respected him, notwithstanding his common abuse of them, by assuring them that they were naturally half beasts and half devils. It was wonderful to see the change soon made in the manners of our inhabitants. From being thoughtless or indifferent about religion, it seem'd as if all the world were growing religious, so that one could not walk thro' the town in an evening without hearing psalms sung in different families of every street.[22]

Even the frugal Franklin was finally persuaded to support the orphanage. In 1740, after once again refusing to contribute, "I happened soon after to attend one of his sermons, in the course of which I perceived he intended to finish with a collection, and I silently resolved he should get nothing from me. I had in my pocket a handful of copper money, three or four silver dollars, and five pistoles in gold. As he proceeded I began to soften, and concluded to give the coppers. Another stroke of his oratory made me ashamed of that, and determined me to give the silver and he finished so admirably, that I emptied my pocket wholly into the collector's dish."[23]

Whitefield's first leave-taking from the city occasioned as much fanfare as had accompanied his arrival. "Many to my knowledge have already been quickened, and awakened to see that religious does not consist in outward things, but righteousness, peace and joy in the holy ghost," he boasted. The news of his coming departure brought even larger crowds to remaining appearances in public and to his apartments. "My house was filled with people, who came in to join in psalms and family prayer . . . their hearts, I believe, were loaded with a sense of sin, the only preparative for the soul-refreshing visitations of Jesus Christ." If this was a return to old strict Calvinism from which even the Church of England had departed, then Calvin had miscounted the proportion of damned and saved. For the door to justification Whitefield opened was as wide as the door to Whitefield's lodgings, and those who poured through the latter believed that the door to the former was as open. Whitefield looked on and was pleased. "Blessed be the Lord for sending me thither."[24]

Then it was time to go. On November 13, Whitefield consigned a packet of letters to be sent to London, where his presence was still felt, and then traveled on by horse to New York City. Franklin spread journalistic fronds along the road north: "Before he returns to England he designs (God willing) to preach the Gospel in every province in America, belonging to the English," or so Whitefield must have confided to Franklin. Franklin asked New York newspaper editor John Peter Zenger to print a poetic notice of Whitefield's ministry. "See! See! He come, the heavenly sound flow from his charming tongue, rebellious men are seized with fear with deep conviction stung. Listening we stand with vast surprise . . . while he declare their crimson guilt . . . while WHITEFIELD to thy sacred strained surprised we silent still . . . and flock around the song."

But even as he asked a fellow printer to join in praise of Whitefield's devotions, Franklin could not curb his own sharp wit—not when it came to organized religion. So he reported this tidbit from London: Whitefield's preaching "is become so offensive to the clergy of this kingdom, that 'tis said one of my Lords the bishops . . . went to the king to desire his majesty to silence him . . . his majesty seemed at a loss how to satisfy the bishop, which a noble duke present observing humbly proposed that in order to prevent Mr. Whitefield's preaching for the future, his majesty would be graciously pleased to make him a bishop." To those upper-class snobs closer to home who "affirm that Mr. Whitefield's tenets are mischievous," Franklin's Plebeian alter ego Obadiah Plainman huffed that they "consider us as a stupid herd, in whom the light of reason is extinguished" and "expect our plaudits by reviling us to our faces," but the throngs of ordinary people who heard Whitefield knew better than this "gibberish." When Franklin took your part, he did not do it by half measures.[25]

Whitefield returned briefly to Philadelphia from the north, this time on his way south. From November 23 to the end of November, he tarried and preached. Though it rained, they stood to hear him, twice a day, as the holy procession made its way south to Annapolis and on to Charles Town, South Carolina (after the Revolution renamed Charleston). In the wake of his second visit, he had left controversy, the contrarian spirit of New England's ministerial fraternity perhaps influencing him to offend more than his usual quota of critics. Richard Peters, rector of Christ Church and a friend of Franklin (Peters was a benefactor of the Library Company and helped found the Philadelphia Academy), openly criticized Whitefield as a ranting zealot. Whitefield replied that Peters was "an entire stranger to the inward spirit." In Charles Town more controversy awaited (the commissary of the Church of England was not pleased, and said so), and more throngs would be coming to hear Whitefield's message.[26]

Whitefield had already sampled Franklin's wry views of religion, apparently, though Franklin made it a policy to be polite. But as persistent as Franklin was in his skepticism, so was Whitefield in ministry. A year after they first met, Whitefield was still at it. The publishing arrangement came first, but it was invariably followed by Whitefield's appeal to Franklin to repent and accept Jesus into his life. "You may print my [biography] as you desire. God willing, I shall correct my two volumes of sermons, and send them the very first opportunity. Pray write to me by every ship, that goes shortly to Charles-Town. . . . Dear Sir, Adieu. I do not despair of your seeing the reasonableness of Christianity. Apply to God; be willing to do the divine will, and you shall know it." Busy still, too. "I think I have been on shore 73 days, and have been enabled to travel upwards of 800 miles, and to preach 170 times, besides very frequent exhortations at private houses." Nor had Whitefield forgotten the ostensible purpose of his mission. "I have collected, in goods and money, upwards of £700 sterling, for the Orphan-house; blessed be God! Great and visible are the fruits of my late, as well as former feeble labours, and people in general seem more eager after the word than ever. O the love of God to Your unworthy friend."[27]

The partnership arrangement benefitted both men, one more evidence, if it were necessary, that both men possessed a keen eye for opportunity. Though it seemed the meeting was a coincidence, in fact it was an almost predictable intersection of the two paths they had traveled. For both had risen from obscurity to fame through circumstances not entirely of their own manufacture.

Franklin Becomes a Printer and Whitefield Becomes a Preacher

IN 1771, in the quiet of an English manor house's garden, Benjamin Franklin looked back on his early life. He could not have predicted that great events lay ahead, for at 65 years of age he was already past the life expectancy of his generation. But his recollection of his days in Philadelphia was crystal clear. By 1740, he remembered, "my business was now continually augmenting, and my circumstances growing daily easier, my newspaper having become very profitable, as being for a time almost the only one in this and the neighboring provinces." Life was good. It had not always been so, however.[1]

Franklin was not a native of Philadelphia. He was born in Boston on January 17, 1706, his father Josiah an immigrant candle maker, his mother, Abiah Folger Franklin, his father's second wife. He was the youngest male offspring of their union. Though his father was prosperous (owning his own home and playing a paying role in his church), even the biggest purse was surely stretched to support twenty children. Benjamin's elder siblings were bound out as apprentices, but young Franklin found little space for himself in what must have still been a very crowded house. His refuge was his reading. He devoured whatever magazines and books he could find or borrow. In 1731, he would help found a Library Company in Philadelphia, so that the young would have books to read and a place to read them.

Franklin had two years of formal schooling and then was apprenticed to his older brother James, a printer. Franklin still scrounged as much time as he could for his reading. He consumed everything from John Bunyan's Puritan allegory *Pilgrim's Progress* to Plutarch's *Parallel Lives* of ancient Greek and Roman statesmen. Though many who could read could not write (writing became common with the introduction of "spellers" later in the century), Franklin was determined to master prose style. He copied from the fashionable magazines of the time, set aside the original, and tried to recall from memory what he had written.

Soon he had the chance to try out his prose. Only fifteen when his brother embarked on a newspaper publishing career with the *New-England Courant*, in 1722 Franklin (now his brother's apprentice) anonymously crafted and delivered to his brother's shop door the "Silence Dogood" essays. Silence was supposedly a middle-aged woman whose life had been one hardship after another. But the voice was pure Franklin, remarkably mature for his tender years, and Silence's self-description fit Franklin's self-image:

> Know then, That I am an Enemy to Vice, and a Friend to Vertue. I am one of an extensive Charity, and a great Forgiver of private Injuries: A hearty Lover of the Clergy and all good Men, and a mortal Enemy to arbitrary Government & unlimited Power. I am naturally very jealous for the Rights and Liberties of my Country; & the least appearance of an Incroachment on those invaluable Privileges, is apt to make my Blood boil exceedingly. I have likewise a natural Inclination to observe and reprove the Faults of others, at which I have an excellent Faculty. I speak this by Way of Warning to all such whose Offences shall come under my Cognizance, for I never intend to wrap my Talent in a Napkin. To be brief; I am courteous and affable, good-humour'd (unless I am first provok'd,) and handsome, and sometimes witty, but always, SIR, Your Friend, and Humble Servant, SILENCE DOGOOD.[2]

As the editorial slant of the paper was anti–Governor Samuel Shute (a smarmy corruptionist), Franklin's satirical commentaries neatly fit its pages, and his brother, unaware of Ben's authorship, published them. There also was more than a little social animosity in the essays, the targets often marked by upper-class hypocrisy and self-interest. At the gate of future prospects, Silence's son William found that "the Passage was kept by two sturdy Porters named Riches and Poverty, and the latter obstinately refused to give Entrance to any who had not first gain'd the Favour of the former; so that I observed, many who came even to the very Gate, were obliged to travel back again as ignorant as they came, for want of this necessary Qualification." But even at the height of his censorship of the unworthy rich, Franklin was no rebel. Wit was his weapon of choice, not rabble-rousing.[3]

Boston was still not ready for such ribaldry. In many ways, it was a city under siege. During the late seventeenth and early eighteenth centuries, war between England and France brought chaos to their northern colonies. From 1689 through 1713 dynastic struggles begun in Europe between England and France and their respective allies came to America as King William's War (1689–1697) and Queen Anne's War (1701–1713). From settlements on Cape Breton Island (Île Royale when it was French), Montreal, and Quebec, French and Indian war parties swooped down on isolated New England towns. From Salem and Boston colonial forces launched counterattacks. Raiding parties on both sides ravaged villages, took hostages, and destroyed crops. Those who could flee the French and their Indian allies found temporary homes in Salem and Boston, bringing with them tales of atrocities. Young Bostonians who joined in the fight often did not return—disease and wounds cost them their lives. Fearful of the disorder and crime of wartime, the city expanded the night watch. Boston was not a happy place. Peace in Europe after 1713 quieted immediate fears, but the generation that came of age in wartime would never forget its terrors. Franklin never did.[4]

Within a year, Franklin's brother discovered the identity of Silence Dogood, and the rivalry between them became intolerable. Ben Franklin left Boston and the remaining years on his apprenticeship behind. Technically, he was a runaway and could be returned forcibly. No attempt was made by his brother or his father to pursue him, however. He made his way to New York City in October 1723 but found that no one wanted a journeyman printer. With no patron or friend in the city, Franklin might have slipped into the ranks of the "able poor," a vagrant assigned to the workhouse. But Franklin did not despair. Though he was not well off, he had a fine store of "social capital"— skill with his tongue and his pen. He was good at making friends. At William Bradford's print shop he learned that William's son, Andrew, might need an assistant.[5]

Andrew's shop was in Philadelphia, and Franklin had enough money to pay in part for the short trip from Manhattan to Perth Amboy, in North Jersey, thence to cross New Jersey to the Delaware River, and so on to Philadelphia. But the boat was caught in a storm, driven away from the Jersey side of the Hudson, and nearly sank off the south shore of Long Island. Franklin was a swimmer, however, and regained the shore. He resumed his trek the next day. Across New Jersey he traveled on foot. Nearly exhausted and feverish, he at last found passage on a Delaware River boat and landed at the edge of Front Street, Philadelphia. "I was in my working dress, my best clothes being to come round by sea. I was dirty from my journey; my pockets were stuffed out with shirts and stockings; I knew no soul, nor where to look for lodging. I was fatigued with traveling, rowing, and want of rest. I was very hungry, and my whole stock of cash consisted of a Dutch dollar and about a shilling in copper." In fact, hardly destitute (his trunk, still on ship, was full of clothing) but still beholden, he gave what he had to "the people of the boat for my passage, who at first refused it . . . but I insisted on their taking it, a man being sometimes more generous when he has but a little money than when he has plenty." Of course, Franklin recollected these events almost fifty years later, when he too had plenty![6]

When Franklin arrived, Philadelphia had two thousand inhabitants. First proprietor William Penn expected his "green country town" to grow and planned the city's dimensions with that in mind. He envisioned a grid pattern that his aide Thomas Holme, a surveyor, designed. The grid pattern facilitated the development of residential and commercial blocks. The original city plot, some 1,200 acres between the Delaware and the Schuylkill Rivers, would become the model for city planning in the future United States. The main streets of Broad and High (now Market) were avenues, and green squares set in the plan resembled those of Christopher Wren's proposal for the rebuilding of London after the Great Fire of 1666. While the Wren Plan did influence the expansion of London to the West, Penn's plan had an unplanned result. Instead of spreading evenly across the land to the west, residents and shop owners clustered up and down Front Street along the Delaware River shore. By the time Whitefield visited, however, the city had over 10,000 residents and was growing toward the west. Years later, Franklin, looking at his city, would find its growth an example of the "American multiplication table."

Walking along the city's swampy piers and pest-ridden wharves, Franklin attributed his own good health to his salubrious habits, in particular his moderation. For many others, Philadelphia was not a particularly healthy

South East Prospect of Philadelphia, ca. 1720 by Peter Cooper. A busy port when Franklin arrived, Philadelphia soon became the largest city in British North America. Library Company of Philadelphia.

place to live. Surrounded by the wetlands of the lower Delaware River and the marshes on the Schuylkill, Philadelphia experienced mosquito-borne malaria and yellow fever in addition to home-grown maladies. The city's population suffered from cholera, smallpox, dysentery, typhoid fever, and other contagious, parasitical, and sexually transmitted diseases. Dampness, population density, poor hygiene, and inadequate diet added to the recipe for early death. Franklin's beloved son died at the age of 4, in 1736, a victim of smallpox and a lesson on the frailty of life that Franklin never forgot.[7]

In the spring and summer of 1744, Annapolis, Maryland, physician Alexander Hamilton toured the northern colonies for his health. His stay in Philadelphia was not particularly pleasant, as he recalled in his *Itinerarium*. His first view of the city came on June 6, 1744. "At my entering the city, I observed the regularity of the streets, but at the same time the majority of the houses mean and low and much decayed, the streets in general not paved, very dirty, and obstructed with rubbish and lumber, but their frequent building excuses that." The weather oppressed him. "The heat in this city is excessive, the sun's rays being reflected with such power from the brick houses and

1. = Market Street Wharf
2. = Crooked Billet Tavern
3. = Thomas Denham's Shop
4. = Andrew Bradford's Printing Shop
5. = Friends Meeting House
6. = The Court House
7. = First Presbyterian Church
8. = Indian King Tavern
9. = Christ Church
10. = John Read's Residence
11. = Tun Tavern
12. = James Logan's Residence

13. = Andrew Hamilton's Residence
14. = Indian Queen Tavern
15. = Prison
16. = Slate Roof House
17. = London Coffee Shop
18. = Pewter Platter Inn
19. = First School
20. = Draw Bridge
21. = Bud's Long Row
22. = Anthony Morris Brewery
23. = Frampton's Brewery
24. = Richard Whitpain's Residence

25. = The Dock
26. = Globe Inn/Ye Coffee Shop
27. = Carpenter's Wharf
28. = Fishbourne's Wharf
29. = Dickinsons's Wharf
30. = Edward Shippen's Residence (the "Governor's House")
31. = Samuel Carpenters Residence

The streets of Philadelphia in the early eighteenth century. Franklin's home and shop on the 300 block of Market Street were not far from the Court House. Map based on contemporary sources courtesy of Billy G. Smith and Alex Schwab.

from the street pavement which is brick." Awnings provided some shade at street level, and balconies above the streets allowed the home owners some respite and perhaps a breeze in the early evening.[8]

For the rest of the day, the city was busy. The shops opened at 5:00 a.m., and the main streets on Penn's grid plan were soon echoing to the clatter of horses and the clang of iron-rimmed wagon wheels. "The market in this city is perhaps the largest in North-America." To it and the shops in the city came the produce of the surrounding farmland—"bread, flower, and pork." Everything not sold in the city was carted down to the Delaware River docks and there loaded on board coasters and ocean-going vessels. Some of the vessels were "outfitting as privateers," bound for the Caribbean to capture French and Spanish merchantmen and bring the goods back to Philadelphia as the spoils of war to be sold at auction or divided among the captain and crew.[9]

When the trade (and the imperial economy) flourished, so did Penn's city. The Quakers led the way. "Shrewd and successful," frugal and honest (in the main), the Quaker merchants profited not only from their own skills but from their close ties with British Quaker families. (Franklin was never a Quaker, but he never dissuaded anyone from mistaking him for one.) Because Philadelphia's mercantile success depended on such ties as the Quakers had with family in the home country, when the British economy suffered reverses, as in the 1720s, Philadelphia suffered. For as Peter Kalm, a Swedish scientist and economist who visited the city in 1748, reported, "Philadelphia carries on a great trade both with the inhabitants of the country and with other parts of the world, especially the West Indies . . . and the various English colonies in North America. Yet none but English ships are allowed to come into this port." The city's fate was tied to the empire's.[10]

It was a city grown hungry, but not yet fat, on the success of its merchants. The Quaker founders of the colony were soon joined by Germans, English, French Huguenots, Scots, and even a handful of Jews. The overseas trade was full of risks—lost ships, lost cargoes, fraud, default, and simply sending the wrong goods to the wrong market. Everything depended on good contacts in England and other ports of call. Trustworthy middlemen were essential, as was cheap warehousing at both ends. Often called a triangular trade (from the mainland colonies, to England, to the Caribbean), the Atlantic trade was far more complex in reality. Philadelphia merchants invested in the slave trade, for example, carrying slaves from Africa to the English West Indies, on-loading sugar and molasses, and returning these to mainland ports.

All Philadelphia shipmasters were bound, at least in theory, to obey the Navigation Acts. These Parliamentary statutes required that certain staple goods produced in the colonies go directly to England, all carriers be of British or colonial registry, and all imports from non-British sources stop first at British ports and be reshipped to the colonies from the home country. As well, the colonies were not to print their own money or to compete with British manufactures. Customs officials in the colonies and royal governors were to police this system. In fact, colonial merchants found ways around the regulations, including smuggling and bribing the customs officials. Everyone in Philadelphia knew who was on the take and who violated the laws.[11]

Merchants and tradesmen often took their meals at the taverns. Hamilton sat among the "Scots, English, Dutch, Germans, and Irish." Hamilton joined in the conversation as it swirled around politics and the war, prices and prospects for trade. He found the evenings' gatherings convivial and the exchange of opinions "agreeable and instructive."[12]

Hamilton barely took note of the servants, laborers, and apprentices who lived on the alleys or the back streets, but the city teemed with them. Philadelphia, once a haven for the oppressed Quakers, had become a warehouse of peoples from all over Europe. "Common laborers" had come from the hinterlands of the colonies, from Wales, England, Scotland, southwestern Germany, Ireland, and ports all over Europe. The very lucky, plucky, or skilled married into the mercantile classes. Some made the leap in status to craftsman, artisan, or shopkeeper. The years of peace before 1739 were good ones for the "middling sort," and prospects abounded for the able, ambitious, and fortunate. Most of the immigrants worked for wages or in apprenticeships for "found" (room and board). Some fell into perpetual poverty, illness or madness stealing away their youth. These destitute men and women found a bed and help at the Almshouse, chartered in 1734. Its location on a rise north of the city was salubrious, but it was little more than a provincial version of the old English poorhouse and a revolving door for a class of marginal men and women.[13]

Slaves filled the streets as well. Franklin owned at least two young male slaves from 1735 until he left for England, in 1757. They were employed in his household and shop. In 1750, he purchased a married slave couple. By 1750, of those free men and women wealthy enough to leave estates for probate in Philadelphia, nearly one-half owned slaves. Gary Nash estimates that fully 15 percent of the dock workers were slaves. Although the Quakers were the first Americans to decry slavery and the slave trade, African bondmen and bondwomen continued to pour into the city. The number would peak during the French and Indian War; some 1,400 out of the 18,000 inhabitants of the city in 1760 were slaves. They worked as house servants, as day laborers, alongside master craftsmen in shops and forges, and in the shipworks. They did the hard and dirty work of cleaning human and animal waste from the streets, alleys, and stables. Some formed families, worshiped in churches, and learned trades. All dreamed of freedom.[14]

Religious diversity (of a limited sort) enlivened the culture of the two cities. Hamilton recorded that "there were Roman Catholicks, Church [of England] men, Presbyterians, Quakers, Newlightmen [evangelicals], Methodists, Seventh day men, Moravians, Anabaptists, and one Jew" at his tavern table. He treated himself to "a very Calvinisticall sermon preached by an old holder" in "whose assembly was a collection of the most curious old fashioned screwed up-faced, both men and women, that ever I saw." At the Roman chapel "I heard some fine musick and saw some pritty ladies. The priest, after saying mass, catechized some children in English and insisted much upon our submitting our reason to religion and believing of every thing that God said (or properly speaking, every thing that the priest says)." As a result, Hamilton "was taken with a sick qualm . . . which I attributed to the gross nonsense proceeding from the mouth of the priest." Philadelphia boasted twenty churches by the end of the French and Indian War, and alongside the Church of England's houses of worship one could find German Reformed, Lutheran, Presbyterian, Congregationalist, Roman Catholic, and a Jewish synagogue.[15]

If the city and Franklin would grow wealthy and wise together, for the present, a teenaged Franklin found himself looking for lodging and work. He dropped by Andrew Bradford's shop and found William, visiting from New York (no doubt his trip not so harrowing as Franklin's). William Bradford introduced Franklin to Samuel Keimer, a printer and shopkeeper. Franklin agreed to Keimer's terms and went to work for him. Though Franklin thought Keimer a "disheveled and quirky man," the older man's penchant for philosophical debate and his odd religious views afforded Franklin many hours of sporting conversation. Always improving the leisure time he found on his hands, Franklin sought friendships with other young men in his situation. In later years, he would turn it into a literary club that he called the "Junto."[16]

In the meantime, Franklin was looking for a patron. The colonial world ran on patronage and clientage—young men seeking to gain the favorable attention of those above them in status, repaying assistance with loyalty. Patronage created dependency. Boston's John Hancock, for example, "made work for people, erected homes he did not need. He built ships that he sold at a loss." But the men whom he patronized repaid him with steadfast political loyalty to him. If one came from the "middling sort," one needed a patron, or so Franklin reasoned. With this in mind, Franklin sought out the lieutenant governor of the colony, Sir William Keith. So well cast was Franklin's written plea to the lieutenant governor for succor that Keith came to Keimer's shop and offered to aid the young man.[17]

Keith helped Franklin return to Boston to visit his family. Franklin took the opportunity to visit with aging Boston Puritan leader Cotton Mather (a target of Silence Dogood), and the two reconciled. Mather was a septuagenarian who was one of the most prolific of American Puritan writers (no less than six hundred sermons and books in print). He still yearned for the time when New England was considered the promised land and Puritans expected Christ to come to it first in the days of final judgment. But Ben and James did not reconcile, and a letter from Keith asking Franklin to return to Philadelphia provided ample excuse for Franklin to bid his family and his birthplace farewell.

Franklin did not tarry long in Philadelphia. He pressed Keith to help him travel to England to learn more of the printer's trade and make useful contacts, what today one would call "networking." Keith offered to supply Franklin with a letter of credit. The letter of credit was the eighteenth-century equivalent of today's bank check, payable to the bearer by the person to whom the letter was addressed. The letter represented funds the author of the letter of credit had supposedly deposited with the person or institution to whom the letter was addressed. Along with other forms of "commercial paper" called bills of exchange and promissory notes, the letter of credit allowed business dealings over long distances among individuals who might not even know one another.

But no letter of credit was ever delivered from Keith to Franklin. Perhaps there had been a misunderstanding. Perhaps Keith had simply failed to honor his promise. There was also the possibility that the letter had been intercepted and another, unintended recipient put the letter to use. When Franklin arrived in London on Christmas Eve, 1724, he had only his wits and will on which to rely.

The nineteen-year-old was dazzled by London, one of the world's capitals, with its elegant Georgian and Augustinian town houses, its ancient guildhalls, and its palaces. He saw as well the twisted alleys that led to the waterfront and the squalor of backstreet rickety wooden walk-ups. Franklin's eye missed nothing. He found work with a printer and for a year plied that trade, avoiding loose women, drink, and infirmity. It was not easy in London to abstain from any of these vices. He moved up from the basement jobs of lugging type to the composing room, saving money, lending money at interest (and collecting), and doing some writing of his own. His talents, judiciously demonstrated, led to a good reputation, something that Franklin was finding essential to success—that and frugality. He returned to Philadelphia a year and a half later—wiser, more cynical perhaps, now a grown man more determined than ever to succeed at his trade and his life.

And he did. He shined as a junior partner in Thomas Denham's retail outlet, as a manager in Samuel Keimer's print shop, as Hugh Meredith's partner in a start-up printing venture, and as the founder in 1727 and leading light of the Junto, a collection of young men who had ambitions (and were of a social station) similar to his own. Somehow he found time to continue his writing, for it truly pleased him to put words to paper.

His *Dissertation on Liberty and Necessity, Pleasure and Pain* (1725), written in the dark London days, explained that the pleasure principle was the driving force in life—not morals, not faith, and certainly not rationality. Raised in a strict Calvinist household whose routines surely included daily prayer, Bible study, and Sabbath church attendance, now Franklin embraced the most radical of deistic language. Perhaps nothing impaired conventional belief so much as tyrannical familiarity. But, and a very big but, who would want to have this unpleasant truth published? "I am sensible that the Doctrine here advanc'd, if it were to be publish'd, would meet with but an indifferent Reception. Mankind naturally and generally love to be flatter'd: Whatever sooths our Pride, and tends to exalt our Species above the rest of the Creation, we are pleas'd with and easily believe, when ungrateful Truths shall be with the utmost Indignation rejected." So he cloaked his sharp view of human nature in sly satires and whimsical pleasantries.[18]

But religion was all around Franklin in this period when pietism was awakening throughout the English-speaking world. Franklin's *Dissertation* may have been part of his resolution of his own identity crisis. Even so, it was extreme for the time, indeed dangerously close to atheism. This being said, his position did not change much over time. He remained an enlightened skeptic. As he wrote to Whitefield in 1764,

> Your frequently repeated Wishes and Prayers for my Eternal as well as temporal Happiness are very obliging. I can only thank you for them, and offer you mine in return. I have my self no Doubts that I shall enjoy as much of both as is proper for me. That Being who gave me Existence, and thro' almost threescore Years has been continually showering his Favours upon me, whose very Chastisements have been Blessings to me, can I doubt that he loves me? And if he loves me, can I doubt that he will go on to take care of me not only here but hereafter?[19]

Skeptical and practical, Franklin turned to experience, not religion, for his moral guide. His advice—frugality, honesty, industriousness, and speaking (in public) no ill—was self-serving, but not insincere. Was he shallow? Not really, for his injunction to be modest was itself immodest, as much a facade as his self-command to be chaste. He was teasing, testing his own ideas by casting them in bold prose. He published his admonitions, made money sharing them, and seems, for the most part, to have lived up to them. After all, in business, a man's reputation was tantamount to his creditworthiness, and Franklin needed credit to engage in his many commercial enterprises. "Credit was the key" to any successful venture that reached beyond the competency of a single household or local trade. Credit opened doors in the other colonies and in England.[20]

In 1729, Franklin turned reputation and credit into a newspaper venture, commencing the publication of the *Pennsylvania Gazette*. Like most colonial newspapers, its four large pages were taken up with news and opinion pieces cribbed from London papers and magazines—no copyright secured or payment offered. Piracy was far more common than copyright permission. Franklin's paper even advertised the arrival of pirated Irish editions of England authors' books. He included announcements of ships' comings and goings, goods imported, local news, and—the specialty of the paper—Franklin's own essays. These he often framed as letters to the editor or contributions from some interested person. The anonymity allowed him to attack rival papers and publishers. The primary victim of these attacks was none other than Andrew Bradford's rival newspaper.[21]

Good contests bring out the voters, and the rivalry boosted sales. But more important was securing two government benefices, the first the publication of official papers (a ready source of income), the second becoming postmaster general of the North American colonies (allowing Franklin to keep Bradford's paper out of the mail pouches of the carriers). The first took some doing, the second some skullduggery, but Franklin had the energy and the wit to accomplish both. He found time to marry Deborah Read, whose affections he had long (if not entirely constantly) courted. A man of means had to have a family. Her husband had deserted her, and her circumstances were in such dire straits that she forgave Franklin an illegitimate son (William Franklin) and reared him with children of their own.

In 1733, Franklin's circle of close friends, the Junto, branched out from conviviality to sponsorship. They founded a subscription Library Company (10 shillings colony money a year bought membership) and pooled their resources to buy books. In the library as well were copies of Franklin's newest venture, *Poor Richard's Almanack*. A new "alter ego," one Richard Saunders, was the pseudonym Franklin chose, aptly because a real Richard Saunders was the compiler of an almanac in seventeenth-century England. Franklin recalled, "I therefore filled all the little spaces that occurred between the remarkable days in the calendar, with proverbial sentences, chiefly such as inculcated industry and frugality. . . . These proverbs, which contained the wisdom of many ages and nations," were actually cribbed from a variety of already published sources. Franklin's readers preferred conventionality to originality if his almanac's sales were any indication, the latter "a sweet source of income for a printer, easily outselling the Bible." Again a rivalry, this time with Titan Leeds, another almanac publisher, promoted sales. The Junto, the *Gazette*, and now the almanac were all sounding boards for Franklin's proposals. A plan for fire companies (based on the principle of mutual assistance) followed, as did a magazine (stillborn) and, in 1737, the long-coveted royal appointment as postmaster general of the colonies.[22]

Affable in public, apparently genial and open, Franklin was also a calculating, ambitious, clever man. Like so many in the eighteenth-century Anglo-American world who aspired to be important and had almost arrived at their destination, Franklin was adept at masking his true feelings. As early as the final epistle from Silence Dogood, he knew that "it often happens, that the most zealous Advocates for any Cause find themselves disappointed in the first Appearance of Success in the Propagation of their Opinion; and the Disappointment appears unavoidable, when their easy Proselytes too suddenly start into Extremes, and are immediately fill'd with Arguments to invalidate their former Practice. This creates a Suspicion in the more considerate Part of Mankind." Poor Richard put it succinctly: "Let all men know thee, but no man know thee thoroughly."[23]

The real Franklin wore masks in public. As sociologist and social historian Richard Sennett has written, "The relationship [between the face behind the mask and the mask that the public sees] is a dyad"—the impression the mask makes is as important as what is behind it. "Appearances in public, no matter how mystifying, still had to be taken seriously, because they might be clues to the person behind the mask." For example, Franklin was never a Quaker, but "for his part, did not neglect to foster this confusion when he shrewdly conceived it to be to his advantage." He simply put on the Quaker mask from his chest of theatrical props.[24]

Such "self-fashioning," in which dress, manner, and mannerisms defined the individual, was not original with Franklin. The English literary tradition into which he bought when he first thumbed through copies of the *Tatler* and the *Spectator* as a youth was established in the late Renaissance. By the eighteenth century, English men of letters had mastered the art of performing an identity. While most visible among those who were courting or at Court, Franklin's self-fashioning was a proof of his own cosmopolitanism and of how the provincial yearned to be part of the metropole.

So who was the man behind the mask that day in November 1739? Not yet the prosperous gentleman of the William Feke portrait of 1748, lace at his neck and wrists, brown wigged, standing in the erect pose of the arriviste; certainly not yet the royal academy Franklin of the 1762 Mason Chamberlain portrait, clothed in plain brown waistcoat and black buckle shoes, sitting at his desk, electrical apparatus hanging behind him while lightning struck the kite outside; in no wise the diplomatic Franklin of Joseph-Siffred Duplessis in 1778, looking out at Paris's elegance with avuncular sympathy. None of these fabrications yet fit him. But he was always eager for new business, and when he heard that Whitefield was coming to town, he knew that opportunity beckoned.

In 1740, twenty-six-year-old George Whitefield wrote a highly unflattering account of his early years. Unlike Franklin's witty, sanitized memoir, Whitefield's was closer to the events, but its solemn and apologetic tone and its admonitory strictures were as contrived in their way as Franklin's writing. Already an ordained minister in the Church of England, Whitefield understood

Benjamin Franklin, by Robert Feke, 1748. Franklin by this time had earned enough money to enjoy an easy life, and the wig suggested his acquired status as a gentleman, but his pose revealed strength and energy, not sloth or self-satisfaction. Harvard University Portrait Collection.

the genre of religious autobiography. Born in sin, suffering in ignorance, the true Christian only came to grace through confession, repentance, and seeking. *A Short Account of God's Dealings with the Reverend Mr. George Whitefield*, published as part of his journals and later in a series of separate volumes, was a narrative of the pilgrim's progress, a sermon in form and highly revealing in content.

Whitefield saw himself chosen for the work at hand, his mission the conquest of the human heart. "A single eye to God's glory" had motivated him to tell his own story. In effect God had moved the pen in his hand, making its words, like the words in the Bible, divinely inspired and Whitefield an instrument of the divine. Were it not for his manifest, almost agonizing sincerity, one might recoil at the author's arrogance.[25]

He was born in Gloucester, in the southwest of England, in December 1714, a difficult birth for his mother, who kept the Bell Inn with his father. The latter died when Whitefield was two. For much of his early childhood he was, in effect, apprenticed to his mother at the Inn. But from a very early age, "stirrings in his heart" told him that he "was born in sin." The pleasure principle dominated his thoughts. "Early acts of uncleanness" gave way to "lying, filthy talking, and foolish jesting." In other words, from a modern perspective he was a normal child—craving attention and acting out to get it. But from the perspective of 1740, of a man born again in Christ, the weight of unnatural sin seemed almost unbearable. "It would be endless to recount the sins of my early days." What is remarkable to the modern reader is that the adult Whitefield seemed so obsessively aware of these sins. But the Puritan who took constant measure of the state of his soul, looking for that faint glimmer of assurance that God had selected him, would not find unusual Whitefield's short account of his unworthiness. "If I trace myself from my cradle to my manhood, I can see nothing in me but a fitness to be damned."[26]

He was not damned, he concluded, for there were also "very early movings of the blessed spirit upon my heart." He could have seen, if at the time he had understood how to look, "the free grace of God" working in him. How to tell the difference between vain hope and real assurance would be the essence of his adult ministry. But that ability only came through a series of steps—recognition that salvation could not come through good works; a sense of his utter helplessness to save himself; rigorous preparation through the study of the Bible and a few interpretive texts; the guidance of John and Charles Wesley, whose "method" taught Whitefield the right steps to take; and then, only then, the blessed assurance that he had been reborn in Christ. As Whitefield would later warn his hearers, "If we once get above our Bibles, and cease making the written word of God our sole rule both as to faith and practice, we shall soon lie open to all manner of delusion, and be in great danger of making shipwreck of faith and a good conscience."[27]

To what extent Whitefield was offering his own pilgrim's progress as a guide to others, and to what extent it reflected actual experience of the child and youth, one cannot tell. If he was as self-absorbed a youth as he wants the reader to believe, surely he would not have experienced his actions as sinful. That could only come with later reflection. In any case, the fact is that his family had suffered a decline in status over three generations and his mother, widowed, had seven children to feed and clothe, and the hardship he experienced then was real.

Young Whitefield recalled that he had a "knack for mimicry and memory for dialogue," at first drawing him to amateur theatricals, and later serving him well in the ministry. He was admitted to a grammar school where he struggled to learn Latin (necessary for college admission) but easily mastered dramatic arts. The well-schooled actor was supposed to use facial and body gestures to convey emotion. At these Whitefield was truly precocious. After a family falling out not unlike Ben's with James, Whitefield left Gloucester to live with an older brother in Bristol. Like Franklin, arriving nearly penniless and unsure of his future, Whitefield found a calling in Bristol. He began to attend church services regularly. He prayed, and an answer came: God intended some great work for him.[28]

There are remarkable parallels between Bristol, Whitefield's adopted home, and Franklin's Philadelphia. Although Bristol was a medieval town and Philadelphia was born at the stroke of William Penn's pen, both were port cities on two rivers whose mouths open wide to the Atlantic: Philadelphia the Delaware and Schuylkill Rivers and Bristol the Avon and Severn. In the early eighteenth century, both cities came to be dominated by merchant elites, the top rank of whom were overseas traders. "The trading spirit permeated the soul of Bristol." Like Philadelphia, "West Indian commerce became the most important branch of [Bristol's] trade." Bristol became the sugar refining capital of the British Isles, and sugar imports provided jobs for a significant portion of the city's laborers and craftsmen. The Bristol merchants avidly joined in the overseas slave trade that made some merchant fortunes in Philadelphia. "Everyone knew that the trade in Negroes was a valuable though risky one." By the 1730s, the Bristol merchants were senior partners in the Guinea slave trade, although Liverpool's ships and merchants eclipsed the Bristol slave trade by midcentury. In addition, a trade in fish, tobacco, and iron products tied Bristol to the colonies. From that trade, Bristol merchants "grew rich" and "her manufactures expanded."[29]

Rapidly changing economic conditions often breed social unrest. Bristol was a case study of this in the 1720s and 1730s. The disparity between the rich and the poor was growing. A new class of laboring poor was drawn to the city by employment opportunities, but the business cycle, dominated by fluctuating market conditions, left the fate of the laboring poor in the hands of others. The razing of three hundred old houses by the docks to build bigger wharves for the merchants took the homes of the poorer residents without recompense. But the hardest hit by the changing economic circumstances were the old crafts—the medieval cloth tradesmen, for example. Apprentices in these crafts were less and less likely to gain a place in the voting lists. Miners, shipyard workers, and others in this new underclass would become ready audiences for Whitefield. His plain message and direct style of speaking appealed to them, as did his willingness to go where they worked and preach. Formerly respected and fully employed families now losing status and income would also find Whitefield's ministry especially attractive.[30]

Whitefield had rejected the opportunities of this rapidly changing material world—neither the craftsman's apron nor the merchant's ledger for him. He found his calling in the Anglican Church, though not yet his voice or his lines. He gained entrance to Pembroke College, at Oxford, as a servitor—fees paid in return for services to the college and its paying matriculates. Seventeen and increasingly convinced of his own weakness (in morals and in physical health, the latter of which plagued him all his life), he plunged into a world of gentlemen and scholars whom those less fortunate, like himself, had to serve. Lonely, he filled his time with "self-imposed religious duties." Charles Wesley and his older brother John, ahead of Whitefield's class at Oxford, asked Whitefield to join them in a Holy Club. At its meetings all sang psalms, prayed, and discussed Bible passages. The group identified a small number of religious tracts that seemed trustworthy guides to Scripture, which Whitefield then memorized. Whitefield found purpose, fulfillment, and ultimately his mission in the "method." Whitefield's personality—aloof but not arrogant, single-minded but not self-serving—complemented the Wesleys' teachings. Only lacking was the venue for an outward expression of his inner transformation.[31]

A "new birth" came to him during one of his episodes of acute mental and physical debility. Home from college, ill, anxious, and needy, he began to preach. Preaching was exhausting but exhilarating. He had found salvation in this world, though salvation in the next was never certain. In 1736 he took his degree and shortly thereafter was ordained in the Church of England. He would never leave that affiliation, though his doctrines and his manner of presentation would cause commissaries, ordinaries, and bishops no end of headache. Ordained at the almost unprecedented age of twenty-two, though not yet fully licensed (that would come in December 1738), he had little chance of finding a pulpit in the metropolitan center.

While he was ministering to people incarcerated in the Oxford jail, a new adventure beckoned. Both John and his younger brother Charles Wesley were carrying on missionary work among the settlers and Indians of the newly established colony of Georgia. Chartered by the crown as a charitable alternative to the workhouse for the honest debtor, Georgia was run like a military camp by its governor general James Oglethorpe. Indeed, the colony was little more than a barrier between South Carolina and Spanish Florida. Oglethorpe and the trustees would allow no slaves and no spirituous liquors in the colony, for both might undermine the colony's security. In 1736, Savannah, like Philadelphia a planned grid city, still looked more like a Roman fortification (Oglethorpe had training as a military engineer and he knew that the Spanish in Florida were watching his every move) than a commercial enterprise. But Savannah was a deep water port and the soil was rich enough to support intensive agriculture. Georgia had a future.

The colony's population, a polyglot collection of Englishmen, German Salzburgers, Scots, and even a few Sephardic Jews, amounted to about three thousand by the end of the proprietary period. Malaria and other endemic diseases, Indian raids, and despair regularly culled these numbers. It was not at first sight an inviting venue for settlement, but perhaps its inhabitants would pay closer attention to the word of God precisely because hardship and danger shrouded the colony.

While Whitefield waited for the chance to continue the Wesleys' mission, he preached in the streets or private dwellings to anyone who would listen. He developed a unique style, appealing to the emotions of his auditory, matching their needs to his neediness. An opportunity to test himself and his approach came in October 1736, when he was asked to temporarily fill a position at the Tower of London. Walking through the London streets in "gown and cassock" (theatrical costuming perhaps still in his thinking), he attracted attention. "See the boy minister," some wag cried out. But the crying began in earnest among his listeners when increasing crowds came to hear him preach. He wept and they followed suit.

He continued as a substitute in other's pulpits until mid-1737, when in Bristol he had the second of his revelations. He realized that his particular message worked best out of doors to a self-selected congregation, people who came to hear him rather than simply to attend church. Back in London in August, he found huge crowds gathering to hear him. To ensure that they could find him, his new convert William Seward placed notices in the London newspapers. Whitefield's ministry had become a phenomenon, the forerunner of evangelical revivalism.

Whitefield's journals recorded his impression of the crowds' responses. Over and over he asserted his belief that they were transformed by his words and his manner. Though the comparison would have appalled him, in some sense he was marketing his performance, an actor on an outdoor stage in a great era of English theater. His purpose was not to divert or entertain, not to win the good will of the crowd, for his message was that the people in front of him were all sinners, and that save for their repentance they would all end up in the fiery pit of hell.

He did not, however, emphasize this fate—that was more like the preaching of his Massachusetts counterpart and soon-to-be ally Jonathan Edwards. Edwards's God justly hated sin and would punish it. As Edwards told churchgoers in Enfield, Connecticut, visiting their church in 1741, "There is no want of power in God to cast wicked men into hell at any moment. Men's hands cannot be strong when God rises up. The strongest have no power to resist him, nor can any deliver out of his hands.—He is not only able to cast wicked men into hell, but he can most easily do it." Whitefield's Calvinism was just as strict as Edwards's—God had chosen the elect before creation; no amount of good works could change one's fate; no evidence of sanctification was proof of justification—but Whitefield emphasized how much God wanted us to seek Him, how much He loved us, how He (and more immediately Whitefield) grieved when we remained stiff necked. One could not will one's own salvation, but one could be born again, indeed, "that I must be born again, and have Christ formed in my heart, before I could have any well-grounded assurance that I was a Christian indeed, or have any solid foundation whereon I might build the superstructure of a truly holy and pious life." This was hardly a promise of eternal life, but there was no chance at all if one were not born again. True faith, found faith, was God's "gift to the believer."[32]

Whitefield's understanding of Calvinist doctrine would deepen some years after he first began his revival preaching. In New England during its Great Awakening he would meet Edwards and other so-called New Lights, see profoundly the difference between their views and those of their critics, align himself with them, and find himself at odds with the Wesleys, Methodism, and the Church of England. All this lay in the future, however. For the present, he was not exploring doctrinal niceties but his own conversion experience, and sharing that with increasingly appreciative audiences.

Whitefield's itinerancy, for he preached on the roads and in the streets and in the fields as well as in churches, without a set pulpit or even a license to one, covered the South of England, from Bristol to London. Everywhere he went, crowds were sure to follow. They asked him to preach, or so he reported (the exact sequence is lost to us, dependent as we are on his journals), and he complied. Rain or shine, his physical debilities always nagging at him, "the doctrine of the New Birth" electrified his audiences.[33]

One is almost tempted to liken Whitefield's tour with that of some modern rock music group, its fans like his converts swooning, crying out for more, save that most often the gathering listened in deep silence broken only by the occasional moan. Unlike the camp meetings of the Second Great Awakening in early nineteenth-century America, there were no "holy rollers," no screams of remorse and fear, not that he reported at least.

As his popularity grew, he found backers for his passage to Georgia. It was a leap from eminence to obscurity. Was it the wilderness that beckoned? Christ found himself in the wilderness, tempted and hardened. Was Whitefield seeking to follow a holy path, away from celebrity, setting self-sacrifice, even danger to his person, as a part of his pilgrimage?

When he boarded ship for Savannah on January 2, 1738, one of his fellow passengers was none other than Oglethorpe. The *Whitaker* was a relatively small ship for its time, barely fifty tons, and the voyage to the colony could take over three months. This was no luxury cruise. The food, even for the better-paying passengers, was unpalatable; the ship was dirty; the sailors were veteran blasphemers; and the sea voyage in winter was at best miserable and at worst deadly. The Wesleys' antislavery message had left behind them a trail of ill feeling and suspicion in Georgia and South Carolina. Although Oglethorpe and the trustees of Georgia had barred slavery, they were anything but abolitionists.

Whitefield worked hard to bring the crew to see the way to salvation, and according to his own account, by the end of the voyage the crew was listening. At Gibraltar, a major British naval base then as now, he preached to his first non-British auditory. There he mingled with Catholics and Jews, the former repulsive to him, but the latter welcoming. He would, throughout his later life, regard the Jews and biblical Judaism with genuine admiration, though he hoped that somehow they would come to see the Truth and convert.

Whitefield gained some allies once he arrived in the colony. Charles Delamotte and James Habersham were Methodists and committed to the orphanage project. Habersham in particular would become the patriarch of one of Georgia's most important families. A merchant by trade, he developed a commercial practice in Savannah to complement his management of Bethesda, Whitefield's orphanage. When Georgia became a royal colony and slavery was permitted, Habersham was ready. His rice plantations would occupy the labor of over two hundred slaves. James died a loyalist in 1775, but his three sons would become revolutionary stalwarts, serve in the army and the confederation congress, and support the federal Constitution of 1787. In the antebellum period, they remained important figures in Georgia's economic and political spheres.

Though he made friends of those the Wesleys had antagonized, conditions in Savannah were rude, and Whitefield's health, never robust, suffered accordingly. But he persevered, reaching out to families, holding conventicles in homes, visiting, singing, praying, teaching, making himself available to everyone. He reversed the Wesleys' views of slavery, agreeing that it should be legal if masters acted as good Christians toward their bondmen. He was not averse to alcoholic spirits either. From his stores on board ship, he carried charitable gifts to the poor in the colony. His reputation grew, but so did his dissatisfaction. He shifted his plan from a permanent ministry to an episodic one, visiting the colony rather than taking up permanent residence, collecting goods and funds all over the colonies and in England and depositing them with trustworthy men of affairs in Georgia, sponsoring churches, schools, and other institutions for the colony while not neglecting the orphanage, and then going on his way. In September 1738, he departed Georgia. He would return periodically to spread his largesse and accept grateful thanks, but never lingering. For, as he wrote to Franklin on June 23, 1747, "If I can say I owe no man anything but love, I do not desire to save a groat, any more than will serve for a visible fund for the Orphan-house after my decease. [That?] institution I cannot give up. . . . If this reasoning be not sound, let me be indulged in the Orphan-house since it is my darling, and a darling in which my own private interest cannot be concerned."[34]

Whitefield did not neglect to publicize his efforts in Georgia. During his stay and after he left, he arranged for the publication of the first of his journals, the centerpiece of which was the journey to Georgia and his stay in the colony. From his vivid account of the storms at sea, to his daily attempts to convert the sailors (the longest part of this first journal), to his detailed narrative of people and places in the colony, he combined religion with travelogue.

The latter had become a very popular genre in the 1720s, including the fabulous and allegorical (for example, Jonathan Swift's 1726 *Gulliver's Travels*) as well as the practical (for example, John Lawson's 1705 *A New Voyage to Carolina*). Lawson's introduction explained the fascination with travel that authors and readers shared: "In the Year 1700, when People flock'd from all Parts of the Christian World, to see the Solemnity of the Grand Jubilee at Rome, my Intention, at that Time, being to travel, I accidentally met with a Gentleman, who had been Abroad, and was very well acquainted with the Ways of Living in both Indies; of whom, having made Enquiry concerning them, he assur'd me, that Carolina was the best Country I could go to." Lawson was also a promoter of the Carolina trade, and again like so many of the travel writers, he had a promotional purpose. Thus, of South Carolina he boasted, "They have a well-disciplin'd Militia; their Horse are most Gentlemen, and well mounted, and the best in America, and may equalize any in other Parts: Their Officers, both Infantry and Cavalry, generally appear in scarlet Mountings, and as rich as in most Regiments belonging to the Crown, which shews the Richness and Grandeur of this Colony"—quite different from slaveless Georgia.[35]

Whitefield too had a promotional scheme in mind. When he returned to London in December, he published the first of the journals. They became a kind of serial, much like the nineteenth-century newspaper serializations of Charles Dickens's novels or Harriet Beecher Stowe's *Uncle Tom's Cabin*, each issue gaining readers, raising the expectation for the next issue. Seward remained indefatigable in his efforts to publicize Whitefield's efforts. Seward "placed two or three notices every week in the *Daily Advertiser*" to ensure that Whitefield's regular followers knew where and when he would preach next. The ostensible purpose was to raise funds for the orphanage. The underlying purpose was to keep his ministry on the front page.[36]

While the pieces were going to press, Whitefield returned to the fields around London to preach. He was not the first to minister in the open air. But he was the first fully ordained and licensed Church of England minister to take to the streets. His "living" or pulpit was to be in Georgia. There was nothing unusual about this. The only licensed Church of England ministers in America were ordained in England and sent to their places in the New World by the bishop of London. Young men sent to England to be ordained sometime fell ill or elected not to return to America. Those who were returned to American parishes sometimes displeased the vestrymen whose tithes paid the ministers' salaries. In the 1760s, this would become a bone of contention in the New England colonies where the Church of England was not the established church. Anglicans in these colonies pleaded for an American bishop to ordain and supervise Anglican ministers. Congregationalists and Presbyterians cried foul.

The London to which Whitefield now devoted his ministry was rapidly changing. It was a city of sharp contrasts, some obvious, some hidden, as Franklin had already discovered. The street life illustrated these contrasts. "A poor man sleeps on a basket lid. An aged crone offers vegetables, while a gallant dallies with two handsome women . . . flying coaches in the background inject movement into a panorama that radiates energy and vitality. . . . Polite customers with money in their pockets have come to buy goods and enjoy the playhouses, alehouses," all of which Whitefield knew from his own experience. But the rich on the street rubbed shoulders with "the fops, gamblers, whores, mendicants, pickpockets," street sharps, and rubes who found their way into the heart of the city. Women who worked in London, for example, in the clothing trades, food preparation industry, and domestic or child care service, joined with the growing number of middle-class women going to and from the shops. Unescorted women were everywhere in the streets, belying the notion that "working women" were somehow disreputable. On the other hand, some "women of the street" were in fact prostitutes. The city attracted the young from the countryside, luring them with the promise of employment and despoiling them with the reality of filth, disease, hunger, and crime. Such contrasts were not new—they were the mark of every world capital. One might have found them in the Rome of Augustus.[37]

London was also the metropolitan center of empire. To it came the Caribbean planters with their slaves, master and servant fleeing from the heat and disease of the Sugar Islands; the children of aristocratic Tidewater Virginians sent to polish their manners and find wealthy wives; Quaker businessmen from Pennsylvania visiting with friends in the trades; and multitudes of ne'er-do-wells, apprentices, minor officials, and the occasional American minister trolling for contributions for a provincial church or school.

London had become the mercantile and banking capital of the world. London merchants, working closely with the Board of Trade, managed a far-flung empire of imports and exports. In consequence, a new stratum of society had appeared. Self-aware and proud, these were the trading classes, the "middle station" and their abettors, the professional men who enabled the economy of the empire to function. This middle class made London its home. Where once the city had been divided among upper-class townhomes and the shanties of the poor, now entire blocks were devoted to homes for the lawyers, doctors, merchants, bankers, and others whose fortunes rose with those of the empire.

In these dwellings material life was more refined than ever before, refined by the purchase and use of consumer durables and early manufactures—the tea and coffee sets, the sitting room furniture and wallpaper, the fine tables and dressers—that defined middle-class life. Even food for the middle class was distinctive and dependent on London's place in the imperial scheme. The coffee, tea, sugar, chocolate, and other imported caffeinates and energy sources kept the middle classes at their desks longer and increased efficiency. Sugar made tea and coffee as popular as alcoholic beverages, and far more likely to keep one awake and busy than beer.[38]

As London city magistrate Henry Fielding wrote of this middle class in 1750, "trade hath indeed given a new face to the whole nation." The successful abandoned "simplicity" for "craft," "frugality" for "luxury," and "humility" for "pride." They were the equal of any man, forgetting the old ordering of society by rank and birth. Such rapid social and economic change in a community often opens a door to religious longing. As the safety and sanctity of old ways is undermined, those who are losing status and those who fear for the future will turn to the reassurance of their faith. Anxiety among the seekers bred the need to find and adhere to evangelical preaching that recognized the spiritual crisis and offered reassurance in it. Such "burned-over" regions then become the birthplaces of new sects or sites for the rejuvenation of older churches in revival enthusiasm.[39]

It was, as events proved, a perfect setting for Whitefield's new brand of ministry. Where once he had preached to the misbegotten masses, now he targeted the market fairs (a forerunner of shopping malls), inns in market centers, and everywhere else that the new class might frequent. He even preached from a shop window to a crowd in the street, literally offering his words as a store-bought commodity. Though he claimed that he was opening his virtual church doors to everyone, in particular commoners who could not sit in the front pews of their neighborhood churches because these were reserved for the wealthy, his target was shifting to those whose pockets held coins. These were the people who could make charitable donations to the Georgia orphanage. They were also a class adrift, without the moorings of the older upper class or the indifference to religion of the lower orders. His successful collections proved that he had found the right audience. The same people were buying copies of the first edition of his journal, cheaply priced at sixpence.[40]

Whitefield's message was also changing somewhat. He was beginning to preach in opposition to other ministers, in particular the more conservative members of his own Anglican order. They leaned too much toward the notion that works and piety led to salvation, a "latitudinarian" persuasion at odds with his Calvinism. These clerics, probably the majority of those in the Church of England, thought that reason combined with Scripture was sufficient to guide the perplexed Christian. They were tolerant of a variety of doctrinal exegeses, as Whitefield, increasingly, was not. They also believed that many, perhaps all, who sought salvation might have been chosen for it, a view that Whitefield found too close to Arminianism, the doctrine that one could will one's way to heaven. Whitefield was also moving away from the Wesley brothers. He kept them as personal friends, but their rejection of predestination (the essential Calvinist principle that God had decreed who would be saved and who damned even before creation) was unacceptable. The formal breach did not occur until the winter of 1741, but it was in the wind. Both brothers had come to believe that all who sought Christ could be saved, as John Wesley explained in "Free Grace," a sermon preached at Bristol in 1740 and later published. Whitefield disagreed. He clung to the theory of "election," that God's chosen were few in number. An exchange of letters followed, and these were published. The public exchange vexed Whitefield. "Why then should we dispute, when there is no probability of convincing? Will it not in the end destroy brotherly love. . . . How glad would the enemies of the Lord be to see us divided?" But he had unleashed the torrent of words; he could not stem the flood now.[41]

For the present, however, all the Methodists seemed united and gloried in Whitefield's success. Seward's newspaper reports of attendance at Whitefield's open-air sessions were exuberant. There were many thousands, he relayed to the publishers, some fifty thousand at one meeting. While the number was exaggerated (Whitefield would lower it considerably in a 1756 edition of his journals), it conveyed to readers the sense that something truly revolutionary was happening. The "staging" of these events took Whitefield back to Bristol, then a return to London, and finally once more to the sea, to Philadelphia and an arrangement with Franklin.[42]

Was Whitefield's piety a mask? Was his overly emotional performance in front of his impromptu congregation an affectation? Was he aware of the impact of his theatrics? Surely yes, and did that heighten for him the need to perform? Again, yes. Franklin saw this almost immediately and at first regarded Whitefield the way a theater critic might regard a great actor in a great role.

What really happened when they met? Who did these two men see in the other when they met? Both performers capable of self-fashioning, did they see in the other a mask maker and wearer? Did they see behind the masks? What did they say to one another after the initial pleasantries and compliments? Were they tentative at first, seeing how different their paths to this place had been? When did they lower their guard, if ever? Franklin was somewhat shy when it came to public meetings; Whitefield was effusive. At some point in the conversation they came to a business understanding. Over time it became more than that, transformed into a genuine appreciation of what was best in each other's character.

The experiences that brought Franklin and Whitefield together in Philadelphia that fall, like the partnership they formed there, centered on the published word. Both men were known not by scepters or swords, but by essays and sermons they wrote and caused to be published. In so far as they fashioned themselves, performing, wearing masks of various kinds, the fabric of the masks was woven of words. In the years to come, the volume of those words would increase in direct proportion to the influence of the two men. Thus, one cannot understand the nature of that influence without a closer attention to their words.

Whitefield's Messages of Hope

three

BOTH WHITEFIELD AND FRANKLIN wrote as they spoke, in a kind of enhanced vernacular. Their language embodied their outlook on the world, and the print media of newspapers and books becoming available to the mass market carried their words throughout the English-speaking world.

Franklin published at least eight volumes of Whitefield's journals covering the period from December 28, 1737, to October 29, 1740. These documented Whitefield's first missionary voyage stop at Gibraltar, his visit to Georgia, his return to London, then back to Georgia via Philadelphia, and then the 1740 trips to New England, to the South, and once more to Philadelphia. The constant motion, across an ocean beset by war, never an easy crossing and often a dangerous one, read like a combination of the seventeenth-century English Puritan classic *Pilgrim's Progress* and a Fodor's Guide to the eighteenth-century Anglo-Atlantic World.[1]

Franklin was not the only publisher to whom Whitefield turned. He found newspaper editors in Boston, New York, and London willing to document his travels and reprint portions of his sermons. The journals already had British publishers. James Hutton had won the competition to publish the first of the journals, buying out rival Thomas Cooper. In their bids for the rights, both Cooper and Hutton had promised that they would sell thousands. Hutton also won Whitefield's blessing to publish the sermons in England. But Franklin was the only publisher of the journals in the New World.[2]

Franklin's publicity for Whitefield's journals took many forms—newspaper squibs tracking the minister's progress through the colonies, depictions of the crowds and their reactions to his preaching, and, most important, advertisements for the books. But the form of these advertisements was different from other newspaper items. Instead of mere facts, they combined rational appeal with hidden persuaders. Franklin even published Whitefield's letters to other ministers, defending himself against their criticism, a correspondence that had become public because his critics had gone public.[3]

Whitefield himself had fired the first salvo in these fusillades, a fact sometimes forgotten. In the December 13, 1739, issue of the *Gazette*, Franklin published a longish letter, supposedly "placed" in the paper by a reader or subscriber. The substance was Whitefield's reputation and the style was pure puffery. Who wrote it? It is possible that Franklin was the author, for he was accustomed to publishing anonymous contributions to the newspaper that he in fact had authored, and he was promoting Whitefield. From his own Calvinist background, Franklin was familiar with Whitefield's language and could duplicate it. The next possibility is William Seward, Whitefield's publicist. Seward was an indefatigable part of Whitefield's entourage during the 1739–1740 colonial tour. But given its similarity in tone, language, and theme to Whitefield's own, one might conclude with reason that he was himself the author of the piece.

The item began "Mr. FRANKLIN, Please to insert the following Extract from the Magazines, in your next Gazette, and you'll oblige many of your Readers." The not-so-hidden message was that Whitefield was already a popular figure, but one whose detractors were loud. The extract did not mention who those detractors might be, but the title, "The Conduct and Doctrine of the Rev. Mr. WHITEFIELD vindicated from the Aspersions and malicious Invectives of his Enemies," suggests that the author was concerned that the animosity Whitefield had already stirred might have an impact on his welcome in the other colonies. The controversy with Richard Peters provided a perfect example and may have occasioned the piece.

Although the subject was religion, the specific content was biographical and the style energetically defensive. "We seem to have in our Day a renew'd

Testimony, that when God has more than ordinary Work to do, he will raise up those who shall be equal to his Purpose; and able to execute all his Pleasure; at the Head of these allow us to place the much misrepresented Mr. Whitefield, the following Account of whom, which we may defy the most malicious of his Adversaries to invalidate, we offer to the World for Truth, in every, even the minutest Circumstance." Whitefield would cite Truth as his goal in sermons and correspondence, but he was not referring to facts. He meant a higher Truth, the true reading of Scripture.

Would Whitefield have blushed to have written the following lines about his own mission? In public, he was self-abasing to the point of parody. But his later letters showed the arrogance of a man certain of his righteousness and his theology. "If ever a Minister of the Gospel endeavour'd to make Christ Jesus his great Example, and tread in the Steps of his immediate Followers, this excellent Person, if we may judge by the whole of his Life and Conversation," was none other than Whitefield. His talents matched his piety. "In Learning not excell'd, if equall'd by any; of superior Abilities to most of his Age; and for a serious Vein of unaffected Piety which accompanies whatever he says or does, he seems to stand alone."

Might the minister have worn a false face? Have practiced to deceive? Have fooled himself as so many did when they preached salvation but were not saved? "God only knows the Heart; but if the Tree is to be known by its Fruit; the humble, the holy, the devout Christian is sure to be seen, whenever he is present; he discovers, upon all Occasions, a thorough Knowledge of, and a large Experience in spiritual Things, to a degree much beyond what could be expected from one twice his Years; and, with a most heavenly Disposition of Mind, discovers the most ardent Affection for Virtue, as well as Religion." Outward piety did not prove that one was a member of the invisible church of the chosen, but Whitefield was special. "He appears animated by an uncommon Zeal for the Glory of God: and not any Person upon Earth can shew a greater Concern for the Good of Souls; his Views seem wholly directed to these; to advance both to his utmost is his Study and Endeavour Night and Day." Whitefield faced criticism that he diverted contributions to the Georgia orphanage to his own use, but his defender contended,

'Twould be doing him great Injustice, at the same Time to deny, that he does not take the utmost Pleasure in Acts of Beneficence and Mercy; in contributing to the Relief of the Poor, according to his own Ability; and in procuring them the Aid of others under their Necessities; In a Word, his whole Life may truely be said, to be one laborious Act of Kindness, on the Behalf of the human Nature; To promote the best, the Eternal Interest of Mankind, and honour his God and Redeemer, he wholly devotes his Time and his Studies, himself, and the Interest he has with others.

The anonymous author next contrasted Whitefield's conduct and demeanor favorably with the mitered and gowned high churchmen in England who doubted Whitefield's sincerity and who would, within a few months, become Whitefield's public targets: "If [his] is not primitive, pure Christianity, if here is not a real Minister of the Gospel, and true Follower of Jesus Christ, where in the World shall we look for, or find either? This is the traduced, the much abused Mr. Whitefield." But in the end he would be known by the effects of his preaching. "Who perhaps, has done more towards reforming a dissolute, vicious Age; and towards bringing the People over, wherever he has come, from their sinful, wicked Courses, to Sobriety and Righteousness of Life; and to the Practice of Religion and Virtue; than, perhaps, half the Clergy in the Kingdom have been able to effect, in ten times the Number of Years."

Why then were the "honest Intentions of his Heart set at nought by the envious Pens of publick Defamers; been the Sport of the Drunkard, a Subject of reproach to men of Hierarchical Principles, and corrupt Minds?" The answer was as plain to the author of the piece as the Truth was to Whitefield. "Was not the Cause of the Disquietude of those last well known, 'twould be no little Difficulty to discover why they are so displeased, or what they carp at." The critics knew that Whitefield lit the true path, the path his critics had departed. While "Mr. Whitefield is against no Man, any farther than such a Man is by Principle or Practice against the People and Interest of his Lord and master," his critics were "so provok'd" by his "popularity and reputation" that they had become jealous and spiteful. But all would be righted, for "the World will be Judge here, between Mr. Whitefield and his Opposers; we doubt not but that within himself he has the Witness of a good Conscience; and we are well assured he has the outward Testimonies of Thousands, of Millions we might say, to his Worth and Desert."

Finally, to those who found in Whitefield's ministry the seeds of incipient democracy, the author of the piece reminded his readers that Whitefield "constantly prays for his Majesty King George, and his Royal Family; with all that are in Authority. He presses Obedience and Submission to the Civil Magistrate; to live peaceable and quiet Lives ... and the Necessity ... to submit themselves, consistent with what they owe to God and their own Souls, to all their Governors, Teachers, Spiritual pastors and Masters, &c. &c." The last few words were especially pregnant with meaning, for ministers and masters in the South were concerned that slaves who had come to hear Whitefield would carry away untoward notions of their own worth and of the wickedness of slavery. Taxed with this charge, Whitefield would scale down his attacks on the inhumanity of slavery and ultimately accept it as a necessary evil.[4]

As one can see from the advertisement, no less than in his acknowledged works, Whitefield, like Franklin, believed that words must be chosen with care. They were the way that he turned Scripture into application. While being born again in Christ might surpass the power of any verbal description, his words urged on that process. His words were his shield against his many detractors, and through words he poured out long and involute replies and defenses.

In 1741, Whitefield took total control of those words. Seward had fallen victim to an angry mob. Whitefield filled the vacant position himself. He supplied London evangelical publicist John Lewis with copy for Lewis's *Weekly History*. Its circulation expanded dramatically to hundreds of subscribers and then, distributed through itinerant evangelical preachers, to another five hundred purchasers. Whitefield sent precise instructions on content and format to Lewis from whatever missionary outpost the minister happened to be visiting. The newly christened *Christian History* was celebratory and optimistic in a fashion that other, earlier venues for Whitefield's writings were not. It also offered Whitefield a forum to defend himself against the growing number of detractors, countering their criticism almost as soon as it appeared.[5]

For all his attempts to shape the good news of his ministry through the journals and newspaper squibs, at the heart of the Whitefield publication efforts were his sermons. He tinkered with the text of them each time he gave one of them, and he revised them for publication. The first of his sermons to be published, "Eternity of Hell Torments," "What Think Ye of Christ?" and "Thankfulness for Mercies Received," all in 1738, are youthful works full of energy and pathos. But perhaps the most affecting and typical of all of Whitefield's sermons was that on the sacrifice of Abraham. He preached it in

Philadelphia during his first visit, and its message and rhetorical rhythms are typical of the rest of his work.

The text was Genesis 22:1–18, one of the most striking and dramatic in all of the Old Testament (or Tanakh, as Jews call their Bible). In it Abraham's faith in God is tested. He is commanded to sacrifice the only son he had with Sarah, the son of his old age, to God. But as Abraham raises his hand to strike Isaac, bound to the rock, the angel of the Lord intercedes and rewards Abraham's faith with the promise of future blessings.

Whitefield's sermons were superb examples of the "plain style" pioneered by the first Puritans. For them, "the guide through the long process was the Word, and the guide through the Word was the sermon." As Whitefield preached in his sermon "Walking with God," "it is every Christian's bounden duty to be guided by the Spirit in conjunction with the written word of God." The sermon was to come "flaming" from the hand of the minister. His contemporary and ally in the Great Awakening, Jonathan Edwards, had elevated the plain style from a somewhat dry and long-winded form of Bible interpretation to an art form. With Edwards and Whitefield, plain style preaching did not mean plain language, but instead avoided "pretentiousness and false sophistication." The plain style was suffused with biblical and literary allusions, metaphors, "puffs of smoke, sudden storms, pieces of food laden with symbolic meaning," and other literary devices to keep the attention of the congregants. "Anecdote, hearsay, and overstatement" were permissible, and "imps, demons, witches, voices in the night, infernal trumpets blaring" brought the invisible world of angels and the devil to light. The plain style allowed the minister to frighten the congregant into paying attention to the lesson of the sermon.[6]

Reading a sermon, however closely, particularly one of Whitefield's sermons, is very different from hearing it delivered aloud. Edwards still read his sermons in the old-fashioned Puritan manner, without gesture or dramatic intonation. Whitefield's version of the plain style shortened the sermon length, and he never read them. Instead, he acted them out as though they were a playwright's scripts. Modern students of Whitefield's rhetoric have no choice but to try to recreate the aural and visual experience from the written word. The format of Whitefield's sermon was not unusual—it is the text-explication-application formula that Protestant ministers had long mastered. The choice of an Old Testament text was also common among the

Puritans. As performed in his unique style, Whitefield's words conveyed pity, torment, love, and hope. Whitefield's dramatization of the text transformed the sermon into a one-man show.[7]

He began quietly, almost studiously, a reminder of the lonely young man's voracious reading habits and the method of exegesis of text that the Wesleys had taught him. He believed that the message in the Bible was complete— one simply had to read intensely and properly to see it. "The great Apostle Paul, in one of his epistles, informs us, that 'whatsoever was written aforetime was written for our learning, that we through patience and comfort of the holy scripture might have hope.'" Assurance of salvation did not come in a blinding revelation, nor could one trust spectral visitations. It was the proper reading of the Bible and other approved theological texts that brought the Christian to the threshold of understanding. To cross that threshold, one had to have faith, to have made a commitment to Jesus. "As without faith it is impossible to please God, or be accepted in Jesus, the Son of his love; we may be assured, that whatever instances of a more than common faith are recorded in the book of God, they were more immediately designed by the Holy Spirit for our learning and imitation, upon whom the ends of the world are come."

The emphasis on faith rather than works was the hallmark of the radical Protestant break with liturgical churches nearly two centuries earlier. No formalistic celebration of sacrament, no outward righteousness, no charitable act was sufficient, for none of these could change the awful judgment of God. But faith mattered. For as Paul told the Hebrews, faith was the meaning of their history. All "the Old Testament saints" were "persons whose faith and patience we are called upon more immediately to follow." They were models for "the Christians of the first, and consequently purest age of the church, to continue steadfast and unmoveable in the profession of their faith."

By recalling a purer past, Whitefield was also recalling a sermon form made popular in New England after the Restoration of the Stuarts in England. "The sins exist, the disease breaks out, the sins are reformed, the disease is cured." This so-called jeremiad called the Puritans back to the piety of their forefathers, much as the Old Testament prophet Jeremiah had. "Amidst this catalogue of saints, methinks the patriarch Abraham shines the brightest, and differs from the others, as one star differeth from another star in glory; for he shone with such distinguished luster, that he was called the 'friend of

God,' the 'father of the faithful'; and those who believe on Christ, are said to be 'sons and daughters of, and to be blessed with, faithful Abraham.'"

Abraham was not a saint in Jewish teachings. He was a patriarch who had made a contract with the Lord, he and his seed to be the Lord's people, the Lord to instruct them in righteous living and keep faith with them after they died. But for Whitefield, Abraham was a "type," a forerunner of Christ, just as biblical Israel was a forerunner of the radical Protestant refuge in America. Whitefield knew exactly where he stood when he preached from the steps of the Philadelphia Court House—on the American shore. Whitefield hinted that Abraham's experience paralleled that of the New World immigrants. "Like many trials of his faith did God send this great and good man, after he had commanded him to get out from his country, and from his kindred, unto a land which he should show him." But Abraham had to do more to prove his faith, by "offering up his only son."

As was his custom when preaching, Whitefield called upon God to be his helper, to give him strength and wisdom. Whitefield believed in a personal God, one who had directed his efforts and guided his travels. "This, by the divine assistance, I propose to make the subject of your present meditation, and, by way of conclusion, to draw some practical inferences, as God shall enable me, from this instructive story." Who could quarrel with his teachings, if God inspired them? Those ministers who did not agree with Whitefield's doctrinal views, or who would not be swayed by his views, must not therefore have divine license to preach. Whitefield would make this point palpably clear when he journeyed to New England during the Great Awakening and there told crowds that his ministerial opponents were unconverted and should be cast from their pulpits.

There were pitfalls in such an invocation of divine communication. When Ann Hutchinson, at her 1638 trial for inciting disobedience to Massachusetts's ministerial fellowship (along with a host of other offenses), told the magistrates that God had spoken to her, they took it as a sign of her antinomian beliefs. Antinomians knew that they were saved, and anyone who doubted it must live in error, teach error, and be damned. How had God spoken to her? Weary at the end of two days of nonstop interrogation in the General Court for allegedly attacking the qualifications of the colony's ministers, she gave the impression of God whispering in her ear, while she probably believed that it had come to her as she read the Bible. The latter was acceptable, the former not. She was already suspected of the former, however, as early as a conversation on her voyage to Massachusetts. Part of the clerical attack on Whitefield was his insistence that he was divinely inspired.[8]

Whitefield's next step was exegesis, the exploration of the meanings of the verses. Today exegesis relocates Scripture in its fullest historical context and employs a wide range of scholarly tools to extract meaning from text. This manner of exegesis was introduced by German Protestant scholars in the nineteenth century. Whitefield's exegesis was based on the belief that the origin of Scripture was divine and thus must be taken literally. Today, this approach would be termed fundamentalist, though that term did not exist until the early twentieth century.

Whitefield, getting down to business: "The sacred penman begins . . . 'And it came to pass, after these things, God did tempt Abraham.' . . . After these things, that is, after he had underwent many severe trials before, after he was old, full of days, and might flatter himself perhaps that the troubles and toils of life were now finished." But no good Calvinist—and Calvinism's stern dictates about human haplessness were the very marrow of Whitefield's preaching—should have rested easy until God's final justification of the chosen. "Christians, you know not what trials you may meet with before you die: notwithstanding you may have suffered, and been tried much already, yet, it may be, a greater measure is still behind, which you are to fill up."

For Whitefield, no sign of election was certain and no appearance of sanctification could be taken on its face. Assurance of salvation, that blessed experience of God's grace, might be false hope in our own ability to save ourselves. "Our last trials, in all probability, will be the greatest: and we can never say our warfare is accomplished, or our trials finished, till we bow down our heads, and give up the ghost." Salvation was God's, not man's, and in the passion of Jesus, the elect (those chosen by God for salvation) are brought to faith, repentance, and a knowledge that they are saved. The saved understand and obey the lessons of the Gospel. "The entire process (election, redemption, regeneration) is the work of God and is by grace alone. Thus God, not man, determines who will be the recipients of the gift of salvation."

There are as many interpretations of Calvin's lessons as there are schools of Protestant religious thought, although for those sectarians who claim to have found the truth of his teachings and through them of God's purpose, there can be no error. The central message of all Calvinist churches is that salvation is by grace, not works, and not human will. Some are predestined to salvation, while the great mass of sinners are damned. Whitefield broke with the Wesleys' teaching that predestination was an error. The Wesleys believed that salvation was utterly dependent on God, but the gates of heaven were open to everyone. "And the children of God may continually observe, how his love leads them on from faith to faith; with what tenderness he watches over their souls; with what care he brings them back if they go astray, and then upholds their going in his path, that their footsteps may not slide. They cannot but observe how unwilling he is to let them go from serving him; and how, notwithstanding the stubbornness of their wills; and the wildness of their passions, he goes on in his work, conquering and to conquer." Thus, for John Wesley, the notion of predestination "utterly overthrows the Scripture doctrine of rewards and punishments."[9]

Wesley and Whitefield agreed that the one book every Christian must study was the Bible. A minister's job included teaching the correct reading of Scripture. "But can the scripture contradict itself?" Why would God tempt a good man to do evil? "Does not the apostle James tell us, 'that God tempts no man'; and God does tempt no man to evil, or on purpose to draw him into sin." Whitefield believed that Satan tempts, and man often slips, for he is born in sin and his sinful proclivities are easily aroused. It would not take much to induce him to act on them. "But in another sense, God may be said to tempt, I mean, to try his servants; and in this sense we are to understand that passage of Matthew, where we are told, that, 'Jesus was led up by the Spirit (the good Spirit) into the wilderness, to be tempted of the devil.'" He resisted, because he knew it was evil.

Whitefield's tone was not fire-and-brimstone, however, but a warm message of comfort. Abraham was everyman, a pilgrim who had found God. Although Genesis did not mention Christ, Whitefield included Him in Abraham's story. It was a kind of license that Christians had taken with the Old Testament, turning the Jewish Bible into a forerunner and part of the Christian Bible. The Gospels started the tradition of connections, citing (with some liberty and discretion) phrases and portions of the Jewish Bible. Jesus himself was Jewish according to the New Testament and was familiar with both the Scripture and the rabbinical commentaries (Talmud) on Tanakh. In any case, Whitefield was following in the path of the gospelers when he implanted Christ in the story of the sacrifice of Isaac.

The gist of the test of faith was God's command to Abraham to "'take now thy son, thine only son Isaac, whom thou lovest, and get thee into the land of

Moriah, and offer him there for a burnt-offering upon one of the mountains which I shall tell thee of.'" This is no ordinary command, for God—Abraham certain what God wanted now—required Abraham to commit a sin. Here Whitefield's reading of the Bible portion differed from Jewish commentary on it. A defining characteristic of biblical Judaism is the total rejection of human sacrifice. For Jews the Akedah, the name given to this portion of Genesis, is all about God's command not to sacrifice anyone. It is not about Abraham's faith, but about God's instructions to the Jews. Hence, it is called "the binding of Isaac."

Whitefield had a different lesson in mind, and to teach it he ceased to be the offstage narrator and became Abraham himself, speaking aloud the patriarch's thoughts. "What! (might the good man have said) butcher my own child! It is contrary to the very law of nature: much more to butcher my dear son Isaac, in whose seed God himself has assured me of a numerous posterity." A good God would not do this. Here was the old and intractable problem of evil. How could a God who loves us cause so much grief to the innocent, the babe who passes away from hunger or disease, the multitudes who die in war and plague, the good men and women like Job who see their earthly treasures cruelly taken from them? The answer was that God had His purposes that a true Christian must accept, however hard it seemed at the time.

Whitefield, still in character as Abraham, worried, "But supposing I could give up my own affections, and be willing to part with him, though I love him so dearly, yet, if I murder him, what will become of God's promise [to multiply Abraham's seed]?" One can imagine Whitefield-the-actor changing his articulation and visage, bending with the weight of Abraham's years as he voiced the old man's inner monologue. Such asides by a character in a play were a regular feature of English drama.

Whitefield then brought Abraham to the American shore, borrowing the language of Matthew 5:14, "a city on a hill," or its explication by John Winthrop, the first governor of Puritan Massachusetts, as he led the great migration of Puritans over the ocean to Massachusetts Bay in 1630: "wee shall be as a Citty upon a Hill, the eies of all people are uppon us; soe that if wee shall deale falsely with our God in this worke wee have undertaken and soe cause him to withdrawe his present help from us, wee shall be made a story and a byword through the world, wee shall open the mouthes of enemies to speake evill of the wayes of God and all professours for Gods sake." So Whitefield had Abraham say, "Besides, I am now like a city built upon a hill; I shine as a light in the world, in the midst of a crooked and perverse generation. How then shall I cause God's name to be blasphemed, how shall I become a by-word among the heathen, if they hear that I have committed a crime which they abhor!"[10]

The lesson was stark and dramatic. "O that unbelievers would learn of faithful Abraham, and believe whatever is revealed from God, though they cannot fully comprehend it!" Good Christians knew that their suffering had a purpose, a divine Providence, which they might not understand but must accept. "Abraham knew God commanded him to offer up his son, and therefore believed, notwithstanding carnal reasoning might suggest many objections." Once again, Whitefield leapt over the millennia, over differences in the language and the customs of peoples to connect the Old with the New Testament. "We have sufficient testimony, that God has spoken to us by his son; why should we not also believe, though many things in the New Testament are above our reason? For, where reason ends, faith begins."

Whitefield not only played the role of Abraham, he effortlessly stepped out of character to compare himself to Abraham. After all, God spoke to both of them. "The humility as well as the piety of the patriarch is observable: he saddled his own ass (great men should be humble)." And both men had averted the snare that even the most human of domestic duties could interfere with obedience to divine commands: "yet he keeps his design as a secret from them all: nay, he does not so much as tell Sarah his wife; for he knew not but she might be a snare unto him in this affair." For Isaac was Sarah's son too, and she had waited as long for him and loved him as much as Abraham, "so Sarah also might persuade Isaac to hide himself."

Women were prominent among Whitefield's converts. He did not look down on them as temptresses or transgressors. His sense of his own masculine identity did not require him to dominate the other sex. He regarded women as candidates for election, as he wrote to one female correspondent in 1749, "I bless God that some have got their faces set Zion-ward, Of the honorable women, ere long, I trust, there will be not a few who will dare to be singularly good, and confess the blessed Jesus before men." But he knew that his revivals might stir "the passions of the weak" unduly, and women were among these weaker vessels of grace.[11]

Abraham, his son, and his servants set out at first light, not knowing where or when to perform the sacrifice. God had told him what to do, but for the rest Abraham must wait. "This was to keep him dependent and watching unto prayer: for there is nothing like being kept waiting upon God; and, if we do, assuredly God will reveal himself unto us yet further in his own time." So the lesson of the portion: wait, prepare, for "what we see not as yet," for "the Lord will reveal even that unto us." One of the oldest quarrels among the Puritans concerned the role of preparation. Did the saints God had chosen have to prepare? According to the ministers, preparation was indeed necessary even for those who believed they had been chosen. But did preparation suffice? Some ministers preached that no amount of preparation would suffice to assure anyone that they were safe in Christ—not without that moment when the light came into their lives.

Ministers had debated, without resolution, whether there were "steps on the way" to justification that the godly must walk, or was that the path of self-delusion that one could will oneself to heaven? Whitefield had read these controversies and knew he had to tread a fine line. The "love of Jesus" was "the only solid preparation for future comfort in the coming world," but this came "experimentally . . . more and more every day."[12]

For the third time, Whitefield entered into the story as Abraham's alter ego. What might seem on its face a dramatic device was in fact a form of exegesis. Abraham was everyman who had faith. By exploring the patriarch's motivations, Whitefield could turn a Bible story into a lesson in psychology. "But who can tell what the aged patriarch felt during these three days? Strong as he was in faith, I am persuaded his bowels often yearned over his dear son Isaac. Methinks I see the good old man walking with his dear child in his hand, and now and then looking upon him, loving him, and then turning aside to weep." Here Whitefield would have wept, as he often did when preaching.

Always ready to assume a role in the story, Whitefield was every personage he described. Every reference applied to him as well as to the biblical cast. The protagonists suffered and he suffered. And when he suffered, his listener suffered. It is called "catharsis," and it is as old as Aristotle's tract *The Poetics*. By means of "pity and terror" we see ourselves as the characters in a staged tragedy. We feel their emotions as our own. In their fall, we fall. Whitefield had some acquaintance with Aristotle and, as a thespian, may well have read *The Poetics*. Whitefield's near-photographic memory exhibited early in his childhood would have enabled him to recall the notion of catharsis.[13]

By now Whitefield's congregation probably stood transfixed, all eyes riveted on him. The silence itself added to the drama. Silently the audience pleaded with Whitefield to go on, to relate what happened next. Then "Abra-ham took the wood of the burnt-offering, and laid it upon Isaac his son; and he took the fire in his hand, and a knife, and they went both of them to-gether." Whitefield became Isaac. "Little did Isaac think that he was to be offered on that very wood which he was carrying upon his shoulders; and therefore Isaac innocently, and with a holy freedom (for good men should not keep their children at too great a distance) spake unto Abraham his fa-ther, and said, 'My father'; and he (with equal affection and holy condescen-sion) said, 'Here am I, thy son.'"

Even the strictest Calvinists were concerned about the state of their chil-dren's souls. In the seventeenth century, they beat the devil out of the child, because children were regarded as miniature adults, sullied by the same original sin, prone to disobey parents and God. But the eighteenth century had introduced notions of a kinder, gentler child rearing in many English Christian households. Discipline was mitigated by sentiment and affection for the child. Following the writings of John Locke and other philosophers, enlightened parents saw children as innocent, their minds an empty tablet on which experience and loving care would write its lessons. In 1739, this transformation was well underway, and Whitefield was part of it. On the one hand, he believed that children came into life with carnal hearts and wicked inclinations, but they were also to be objects of real affection. Parents should treasure them. For "how careful Abraham had been (as all Christian parents ought to do) to instruct his Isaac how to sacrifice to God." The very meta-phor of salvation was the innocence of childhood. "Are ye converted? Are ye become like little children? . . . If ye are converted and become like little children, then behave as little children. . . . Verily I say to you, except ye be converted and become as little children, ye shall not enter into the kingdom of heaven." Love God unhesitatingly, perfectly, "then grow in grace, and in the knowledge of your lord and savior Jesus Christ."[14]

Many in Whitefield's congregation were young men and women. They would become the core audience in the Great Awakening in the New World. Whitefield saw himself in them. "How beautiful is early piety! How amiable, to hear young people ask questions about sacrificing to God in an acceptable way! Isaac knew very well that a lamb was wanting, and that a lamb was nec-essary for a proper sacrifice." What did Isaac know of his fate, and when did he know it? It was the fate of all God's creations to die and then to be judged. For the Calvinist, judgment had occurred before creation itself. How stern was this judgment? Was it according to man's merits? Whitefield believed in predestination, but this did not mean one's conduct and thoughts went unnoticed. God's decree might come before creation, but Whitefield's God was an immanent God, never far away from his errant human creations. So Abraham's God watched over him. Isaac's God watched over him. Was Isaac's painful and early death then to be a punishment for his sins? No. God knew that Isaac was a dutiful and pious son. So did Abraham. "And Abraham said, 'My son, God will provide himself a Lamb for a burnt-offering.'"

Abraham was honest as every parent must be about death and judgment with his or her children. "Whatever Abraham might intend, I cannot but think he here made an application, and acquainted his son, of God's dealing with his soul; and at length, with tears in his eyes, and the utmost affection in his heart, cried out, 'Thou art to be the lamb, my Son.'" Isaac did not resist, for his father only carried out God's will. "He was partaker of the like precious faith with his aged father, and therefore is as willing to be offered, as Abraham is to offer him."

The emotional level of Whitefield's retelling had risen along with the emotional level of the tale, and such great emotion could not be carried for too long. A master at manipulating the emotions of his listeners, Whitefield knew they needed to catch their breath. "And here let us pause a while, and by faith take a view of the place where the father has laid him." Like the bundle of firewood on which Abraham had laid Isaac, colonial Philadelphia was a tinderbox. The potash works, the blacksmiths' shops, and the barns where hay and lanterns lay next to one another all posed the everyday risk of catastrophic fires, with life expectancies dipping in the period from 1720 to 1740. In the old world and the new, the heaviest burden fell on the very young and the very old, whose immune systems could not cope with the wide variety of maladies. Doctors were little help as the germ theory of disease lay in the future. "Cunning folk" selling magical and herbal remedies probably were less dangerous than the surgeons with their dirty lancets. "Come, all ye tender hearted parents, who know what it is to look over a dying child" was no fanciful image. Parents could expect one child in two to die before a fifth birthday.[15]

Parents and children had to be ready to die. "Methinks I see the tears trickle down the Patriarch Abraham's cheeks; and out of the abundance of the heart, he cries, Adieu, adieu, my son; the Lord gave thee to me, and the Lord calls thee away; blessed be the name of the Lord." At this dramatic moment, the crux of the story, Whitefield dropped the mask of thespian and became the minister. "But why do I attempt to describe what either son or father felt? It is impossible: we may indeed form some faint idea of, but shall never full comprehend it, till we come and sit down with them in the kingdom of heaven, and hear them tell the pleasing story over again."

For the story was, in Whitefield's hands, only a means to an end—the reform of the sinner. And atonement lay not in Whitefield's power, no more than in the sinner's, but in God's. God wanted men to reform. "But sing, O heavens! and rejoice, O earth! Man's extremity is God's opportunity." Whitefield reported that God's angel stayed Abraham's hand just as his knife was at Isaac's throat. "'And he said, Lay not thine hand upon the lad, neither do thou any thing unto him: for now know I that thou fearest God, seeing thou hast not withheld thy son, thine only son from me.'" Abraham had passed the test. "Now it was that Abraham's faith, being tried, was found more precious than gold purified seven times in the fire. Now as a reward of grace, though not of debt, for this signal act of obedience, by an oath, God gives and confirms the promise, 'that in his seed all the nations of the earth should be blessed.'"

The application the members of the audience took from his words was no doubt different according to the beliefs of the listener. The unconverted heard Whitefield say that God's promise of eternal life was for all who had faith. "With what triumph is he now exulting in the paradise of God, and adoring rich, free, distinguishing, electing, everlasting love, which alone made him to differ from the rest of mankind, and rendered him worthy of that title which he will have so long as the sun and the moon endure, 'The Father of the faithful!'" The last reference might be confusing to the unconverted, but the elect in the crowd knew that with those five words Whitefield was speaking just to them. The reference to "electing, everlasting love" reminded the Calvinists in Whitefield's ministerial audience that he was a strict Calvinist.

Whitefield, looking out over a sea of faces, knew what effect his words had on the unconverted. "I see your hearts affected, I see your eyes weep. (And indeed, who can refrain weeping at the relation of such a story?) But, behold, I show you a mystery, hid under the sacrifice of Abraham's only son." The mystery was easily solved. Only the "love of God, in giving Jesus Christ to die for our sins," could save. The binding of Isaac was a precursor, a parallel. But for whom was Jesus's sacrifice? Isaac's sacrifice was to test Abraham's faith, but Abraham was powerless to affect his or his son's future. All men were helpless. Yet in their love of Jesus, reciprocated, there was help. "Yes;

that is it. And yet perhaps you find your hearts, at the mentioning of this, not so much affected. Let this convince you, that we are all fallen creatures, and that we do not love God or Christ as we ought to do: for, if you admire Abraham offering up his Isaac, how much more ought you to extol, magnify and adore the love of God, who so loved the world, as to give his only begotten Son Christ Jesus our Lord, 'that whosoever believeth on Him should not perish, but have everlasting life'?" And the payoff for coming to Christ with an open and willing heart? What was that if one's fate in the hereafter was already sealed? What did conversion gain one?

Now came Whitefield's last words, building to a climax, linking Old Testament to New, and thence to Whitefield's listeners: "A ram is offered up in Isaac's room, but Jesus has no substitute; Jesus must bleed, Jesus must die; God the Father provided this Lamb for himself from all eternity. He must be offered in time, or man must be damned for evermore." To the throng, now hanging on Whitefield's every word and gesture, Jesus suffered for the sinners' refusal to repent. "And now, where are your tears? Shall I say, refrain your voice from weeping? No; rather let me exhort you to look to him whom you have pierced, and mourn, as a woman mourneth for her first-born: for we have been the betrayers, we have been the murderers of this Lord of glory." But the deed could be undone. "Having so much done, so much suffered for us, so much forgiven, shall we not love much! O! let us love Him with all our hearts, and minds, and strength, and glorify him in our souls and bodies, for they are his."

What will save the Christian? What will avail the sacrifice of Jesus? The lesson starts out as Calvinism: "Whoever understands and preaches the truth, as it is in Jesus, must acknowledge, that salvation is God's free gift, and that we are saved, not by any or all the works of righteousness which we have done or can do: no; we can neither wholly nor in part justify ourselves in the light of God." Good works cannot save. The ministrations of the holy cannot save. The sacraments of the church cannot save. "The Lord Jesus Christ is our righteousness; and if we are accepted with God, it must be only in and through the personal righteousness, the active and passive obedience, of Jesus Christ his beloved Son." But here Whitefield departed from the Puritans of older days and edged away from strict Calvinism. He did not stress false consciousness, self-deluded sanctification. Instead, he offered hope. "That very moment a sinner is enabled to lay hold on Christ's righteousness by faith, he is freely justified from all his sins, and shall never enter into condemnation, notwithstanding he was a fire-brand of hell before."

It is not that perfect faith is a sign of assurance, a proof that God had chosen the sinner to be saved. For one can still fool oneself. "You say you believe; you talk of free grace and free justification: you do well; the devils also believe and tremble." But "has the faith, which you pretend to, influenced your hearts, renewed your souls, and, like Abraham's, worked by love? Are your affections, like his, set on things above? Are you heavenly-minded, and like him, do you confess yourselves strangers and pilgrims on the earth? In short, has your faith enabled you to overcome the world, and strengthened you to give up your Isaacs, your laughter, your most beloved lusts, friends, pleasures, and profits for God? If so, take the comfort of it." What comfort? "We know assuredly, that we do fear and love God, or rather are loved of him."

Whitefield warned that a time had come when life was so precarious that the true Christian must reform without delay. "Learn, O saints! From what has been said, to sit loose to all your worldly comforts; and stand ready prepared to part with everything, when God shall require it at your hand." But it was possible for everyone in the auditory to strive toward that sainthood. "And as for you that have been in any measure tried like unto [Abraham] let his example encourage and comfort you." Everyone might be among the elect. "Remember, it will be but a little while, and you shall sit with them also, and tell one another what God has done for your souls. There I hope to sit with you, and hear this story of his offering up his Son from his own mouth, and to praise the Lamb that sitteth upon the throne, for what he hath done for all or souls, for ever and ever."

Wherein lay the difference between Methodism and more traditional Calvinist views? The answer lay in the words of the sermon itself: "has the faith, which you pretend to, influenced your hearts, renewed your souls?" Have you been reborn? "The new birth, or change of heart" wrought by the congregant's seeking faith, urged on by the minister, the sermon becoming the vehicle and the revival meeting the occasion, was the difference. "This foundation, as well as this sudden and instantaneous change," happened in real time to real people, not in the timelessness before creation, in God's time. "I mean receiving a principle of new life, imparted to our hearts by the holy ghost, changing you, giving you new thoughts, new words, new actions, new views so that old things pass away, and all things become new in our souls.

. . . When we speak of a new birth, we do not speak of a cunningly devised fable; what our eyes have seen, our hands have handled, and what our hearts have felt." One can trust this new birth, as the old Puritan could never fully trust an experience of grace. The Holy Ghost will not fool Whitefield or those like him who have experienced a new birth.[16]

Men are helpless to save themselves but not helpless to find God. They are wretched creatures prone to sin, but they can be better than that. They can improve themselves. Men may not have the power to gain Heaven through works or piety, but they can find themselves in Christ. Whitefield's message was one of individual empowerment.

It was also a message about human happiness. Wretched though the sinner might be, he could find happiness in Christ—not just grace, some inkling of God's love, but genuine happiness. It was this happiness that sustained Whitefield in his times of weakness, enabled him to preach three times a day, to travel stormy seas and rough dirt roads, to love even those who denounced him. He was happy, and so might be those who were born again in Christ.

Above all, then, the message was one of hope. When Whitefield met Jonathan Edwards in Northampton, Massachusetts, on October 18, 1740, he recalled in his journal, "I came into his pulpit, I found my heart drawn out to talk of scarce anything beside the consolations and privileges of saints, and the plentiful effusion of the spirit upon believers." The next day, preaching, Whitefield "had an affecting prospect of the glories of the upper world." He had begun, like the sinner, "with fear and trembling," but had come, as he hoped his listener would, to a "refreshing," in the presence of the Lord. That was what he wished for all Christians—hope, happiness, and a sense of peace.[17]

All this in one sermon (and sometimes preaching three of them a day)—no wonder that Whitefield complained of weakness. For nearly thirty-two years in America, from 1738 to 1770, on land and aboard ship, in fields, churches, and private homes, Whitefield preached. He found time to console, sing, and pray with others. He wrote as well, composing letters, tracts, journals, and versions of his sermons.

The extant collections of his letters represent only a tiny portion of the whole, but his replies to individual correspondents were like mini-sermons, reassuring, inspiring, and comforting. "Ere long we shall all arrive, I trust, in Abraham's harbor; from thence we shall never put out to sea any more. There the wicked world, and even God's own children, will cease from troubling, and our weary souls enjoy an everlasting rest! May you and yours enter with a full gale!" Other letters were expanded thank-you notes: "My poor prayers shall not be wanting, in your behalf. That is the only return I can make to you both, for the great kindness conferred on me at your house." But the bulk of the preserved correspondence was didactic—Whitefield battling his critics. For example, after he had published his reservations about the Moravians and their leader, Count Zinzendorf, whose views of universal salvation and ecumenical harmony Whitefield rejected, he continued, "Your writing in such a manner convinces me more and more that Moravianism leads us to break through the most sacred ties of nature, friendship, and disinterested love. . . . O my dear man, let me tell thee, that the God of truth and love hates lies, and that cause can never be good, which needs equivocations and falsehoods to support it," the lies being those supposedly told by Whitefield's enemies.[18]

His long and tortured correspondence with John Wesley was an example of the latter kind of correspondence. After Wesley had preached that the door to heaven was open to all, not merely those predestined to pass through its gates, Whitefield went out of his way to correct his old mentor. As he wrote to Wesley on December 24, 1740, "For God is no respecter of persons, upon the account of any outward condition or circumstance in life whatever, nor does the doctrine of election suppose him to be so. But as the sovereign lord of all, who is debtor to none, He has a right to do with his own, and to dispense his favors to what objects He sees fit, merely at his pleasure." Let the sinner repent, the convert seek, the Christian believe—it was all no matter, for the elect were named by God before creation. "A certain number was then given," and only for these did Christ die. The letter—a letter to a friend—reveals a man never totally at rest, driving himself from place to place and never seeming to find peace in any one of them though he preached peace everyplace he went.[19]

In the light of Whitefield's doctrinal rigor as expressed to Wesley, was the promise of God's love and the comfort of conversion in the Sacrifice of Abraham then nothing more than a preacher's trick to conjure conversion? Was the promise that anyone who sought Christ might find in Him the way to salvation a cunning lie? Surely some Calvinist preachers engaged in this practice, knowing that true assurance was only there for those chosen few. Might one even suggest that by engaging in such a ploy Whitefield was something of a hypocrite? Or perhaps a better reading of the man was this: in front of those seeking a way to heaven through works and faith, Whitefield's stern doctrines melted, melted with his heart and his tears, and he spoke his hopes for them rather than his fears.

Whitefield believed that every man was tempted by the implacable ire of Satan. Like Jesus himself, he felt the gall of Satan pour over him. But in Jesus rejecting Satan might not anyone who sought Jesus find his succor? "We may therefore from this blessed passage [from Matthew 4:1–11] gather strong consolations; since by our Lord's conquest over Satan, we are thereby assured of our own, and in the mean while can apply to him as a compassionate High Priest, who was in all things tempted as we are, that he might experimentally be enabled to succor us when we are tempted." Even those who had walked in sin, scoffed and snickered, might place their feet upon the right path before they stumbled into eternal damnation. "Welcome Jesus, welcome thy word, welcome thy ordinances, welcome thy Spirit, welcome thy people, I will henceforth walk with you. O that there may be in you such a mind! God will set his almighty fiat to it, and seal it with the broad seal of heaven, even the signet of his holy Spirit. Yes, he will, though you have been walking with, and following after, the devices and desires of your desperately wicked hearts ever since you have been born." God "will set" (not had already set) his fiat upon their fate if the sinners convert in time. The Word of the Lord would set the sinner free. As Whitefield wrote to Franklin on February 26, 1750, "As we are all creatures of a day; as our whole life is but one small point between two eternities, it is reasonable to suppose, that the grand end of every Christian institution for forming tender minds, should be to convince them of their natural depravity, of the means of recovering out of it, and of the necessity of preparing for the enjoyment of the supreme Being in a future state. These are the grand points in which Christianity centers."[20]

But Whitefield's words were beginning to trap him in a web of controversy, as his public and private exchanges with other ministers grew hot. Whitefield was a different man, less loving and giving, when he donned the gown of the controversialist, as his role in the colonial Great Awakening would prove.

Franklin's Essays on Improvement

FRANKLIN LOVED WORDS. He loved to play with them. If he were alive today, he would have his own blog. Poor Richard and the *Gazette* served a similar purpose for him, a way to "post" his opinions on a variety of topics, some current, some more general. When he was apprenticed to his brother, he did not have much in the way of property or prospects (though his family was far from destitute), but words he could own. From the time that he discovered how much impact they could have in his Silence Dogood essays, he made words the center of his intellectual endeavors. As a printer and then a publisher, he learned that words were fungible. One could set movable type in their trays in different arrays. Words also paid; with this realization, his sharp business eye led him to the prospects of Whitefield's sermons and journals.

Franklin did not see words as Whitefield did, as a link to a divine source. He did not believe in the literal truth of the Bible. How different this was from Whitefield's preaching. But Franklin did preach, after a fashion, his own sermons. These were essays on improvement of the world, rather than the soul.

Franklin's essays had both a playful and a serious side, as his admonitions to himself on religion displayed. Poor Richard poked fun, but there was a kernel of wisdom in Poor Richard's sayings as well. Some of Franklin's essays, for example, on the militia or the need for fire companies, were highly practical. Other essays, such as those on the need for a Philosophical (i.e., scientific) Society, were forward-looking and purposeful. Though Franklin in person was invariably polite to the point of circumspection, Franklin's printed words could do battle, particularly with rival almanac publishers and newspaper printers. His war of words with printer and publisher Andrew Bradford was a classic example.

Before launching his own newspaper at the end of 1729, Franklin took a very sly swipe or two at Bradford's newspaper the *American Weekly Mercury* in a series of letters from one "Busy-Body." Bradford printed these in February and March 1729, not quite realizing who had written them or what Franklin had planned. The first of them set the tone: "I have often observ'd with Concern, that your Mercury is not always equally entertaining. The Delay of Ships expected in, and want of fresh Advices from Europe, make it frequently very Dull; and I find the Freezing of our River has the same Effect on News as on Trade." Lest Bradford guess that Franklin's purpose was less than complimentary, Franklin shifted to complaints on more general topics: "With more Concern have I continually observ'd the growing Vices and Follies of my Country-folk. And tho' Reformation is properly the concern of every Man; that is, Every one ought to mend One; yet 'tis too true in this Case, that what is every Body's Business is no Body's Business, and the Business is done accordingly. I, therefore, upon mature Deliberation, think fit to take no Body's Business wholly into my own Hands." Two puns, a little twisting of words, all for fun, all showing Franklin's love of language, yet there was a moral purpose to his punning. He cloaked his criticisms with self-depreciation, at once modest and sharp: "out of Zeal for the Publick Good, design to erect my Self into a Kind of Censor Morum; proposing with your Allowance, to make Use of the Weekly Mercury as a Vehicle in which my Remonstrances shall be convey'd to the World."[1]

Later in 1729 Franklin explained his editorial policies for the *Gazette* in a note from "The Printer to the Reader." He mixed the wry and the serious in more equal measure than he had in the Busy-Body essays. "There are many who have long desired to see a good NewsPaper in Pennsylvania; and we hope those Gentlemen who are able, will contribute towards the making This such." Wink, wink, nod, nod—another slap at Bradford was concealed

within the outwardly sober call for publishable contributions and reader subscriptions. The announcement masked Franklin's true intent. "We ask Assistance, because we are fully sensible, that to publish a good News-Paper is not
so easy an Undertaking as many People imagine it to be." What was needed?
Nothing more or less than a mastery of words. "The Author of a Gazette (in
the Opinion of the Learned) ought to be qualified with an extensive Acquaintance with Languages, a great Easiness and Command of Writing and Relating Things cleanly and intelligibly, and in few Words." And who could supply this need? None other than Franklin himself. Then, lest his pride show,
Franklin qualified his claims: "Men thus accomplish'd are very rare in this
remote Part of the World; and it would be well if the Writer of these Papers
could make up among his Friends what is wanting in himself."[2]

Franklin was the author of most of the original pieces in the *Gazette*. Some
of these were fictitious, and one of the most brilliant of the latter, typical of
the moralism at the core of Franklin's wit, was the report of a witchcraft trial
in Mount Holly, Burlington County, New Jersey.

In pillorying the fictitious New Jersey events, Franklin was referring to the
superstitions that underlay the actual witchcraft trials in his own birthplace
of Massachusetts. Anyone reared in New England knew about the horrific
excesses of the Salem witchcraft trials of 1692. Then nineteen women and
men were tried in special courts and executed for consorting with the devil
and borrowing his powers to prick with pins and bruise a number of pre-
adolescent and teenage girls, turn cheese and butter bad, frighten babies to
death, and otherwise cause mayhem. Another defendant was pressed to death
with heavy stones for refusing to plead to the charges. The evidence against
the accused was composed of malicious rumor and "spectral" visitations of
the witches only visible to the victims. Nearly two hundred more accused
waited in prison for their turn at trial. By the time regular courts were called
to hear the rest of the cases, in January 1693, judges and juries had come to
their senses, spectral evidence was excluded, and all defendants were either
not indicted, acquitted at trial, or pardoned by the governor.

Superstition was always one of Franklin's favorite targets, and when religion and superstition held hands, his abilities as a satirist equaled those of
contemporaries such as Daniel Defoe and Jonathan Swift. It hardly seems
fair to interrupt Franklin's account. With some elisions and editorial comments of my own, here it is: "Burlington, Oct. 12. Saturday last at Mount-
Holly, about 8 Miles from this Place, near 300 People were gathered together

to see an Experiment or two tried on some Persons accused of Witchcraft." The oblique reference is to the "tests" by which medieval English authorities determined if the accused was a witch. These "trials by ordeal" included scalding some portion of the suspect's body and seeing if it healed. If it did not, the accused was a witch. Dunking the alleged witch in a pond, to see if she floated (thus a witch), was another common ordeal. The entire community turned out for such ordeals. "It seems the Accused had been charged with making their Neighbours['] Sheep dance in an uncommon Manner, and with causing Hogs to speak, and sing Psalms, &c. to the great Terror and Amazement of the King's good and peaceable Subjects in this Province." Franklin made the specific evidence of the malice of the witches so ludicrous that even the most gullible reader would get the joke.

But in Salem there had been no joking. Authorities took the accusations very seriously. So Franklin took his next swipe at the witch finders. "The Accusers being very positive that if the Accused were weighed in Scales against a Bible, the Bible would prove too heavy for them." Supposedly, witches would get tongue-tied if they tried to recite certain passages from the Bible. Why the devil, who gave them the power to fly through the night, to turn themselves into invisible specters and attack the innocent, and to cause illness and death, could not also enable them to recite the Bible perfectly from memory was never explained. (Indeed, why they could not use their powers to assault the judges or the jurors at trial was never explained.) "Accordingly the Time and Place was agreed on, and advertised about the Country; The Accusers were 1 Man and 1 Woman; and the Accused the same." The trial by ordeal had become a tag team wrestling match.

Next, "a Committee of Men were appointed to search the Men, and a Committee of Women to search the Women, to see if they had any Thing of Weight about them, particularly Pins." In Salem, midwives examined the accused women to see if they had any "devil's marks" on their bodies. The reference to pins is a sly aside about the afflicted girls' testimony in Salem. They hid pins in their clothing and pricked one another to convince the examining magistrates that the witches's specters were pricking them.

Almost all the accused in Salem were lower or middle class. The better sort were excused from the exercise—which made them the next target of Franklin's satire. "After the Scrutiny was over, a huge great Bible belonging to the Justice of the Place was provided, and a Lane through the Populace was made from the Justices House to the Scales, which were fixed on a Gallows erected for that Purpose opposite to the House, that the Justice's Wife and the rest of the Ladies might see the Trial, without coming amongst the Mob."

At Salem, in fear for their lives, some of the defendants confessed that a tall dark man had come among them and made them sign the devil's book in their blood. Franklin had probably read Robert Calef's stinging contemporary account of the trials, borrowing its details in his parody. "Then came out of the House a grave tall Man carrying the Holy Writ before the supposed Wizard, &c. (as solemnly as the Sword-bearer of London before the Lord Mayor) the Wizard was first put in the Scale, and over him was read a Chapter out of the Books of Moses." On the gallows at Salem, before they were executed, the convicts were preached at and prayed over by ministers. One of them was the very same Cotton Mather (a great believer in the power of the devil, the existence of witches, and the value of the witchcraft trials to scourge the land) whom Franklin had attacked in his Silence Dogood essays.

Finally, the test: "and then the Bible was put in the other Scale, (which being kept down before) was immediately let go; but to the great Surprize of the Spectators, Flesh and Bones came down plump, and outweighed that great good Book by abundance. After the same Manner, the others were served, and their Lumps of Mortality severally were too heavy for Moses and all the Prophets and Apostles."[3]

The Salem witchcraft trials were never again duplicated in the colonies. Witchcraft accusations had not gone out of fashion, but magistrates refused to take them seriously. Public drunkenness was a social problem that could not be ignored, however. Attempts by legislatures to limit the number of taverns and inns serving alcohol were evaded or ignored in the same fashion as the colonists generally ignored English customs laws or bribed customs officials. In an age when alcoholic beverages were served to people of all ranks and ages, men and women, and public drunkenness was the most common offense on the justice of the peace's docket, Franklin was an advocate of temperance. He did not drink. While not condemning those who did outright, he found the right words to show his views by collecting all the terms to describe a drunk.[4]

So-called devil's dictionaries were a popular form of literature in Franklin's day. His drinker's dictionary is a tour de force. "Tho' every one may possibly recollect a Dozen at least of the Expressions us'd on this Occasion, yet I think no one who has not much frequented Taverns would imagine the number of them so great as it really is. It may therefore surprize as well as divert the sober Reader, to have the Sight of a new Piece, lately communicated to me, entitled The Drinkers Dictionary." The dictionary (of course his own compilation) listed, among other entries,

> Bewitch'd, Boozy, Bowz'd, Been at Barbadoes, Piss'd in the Brook, Drunk as a Wheel-Barrow, Burdock'd, As Drunk as a Beggar, Head with Sampson's Jawbone, Cagrin'd, Cherry Merry, Wamble Crop'd, Crack'd, Half Way to Concord, Got Corns in his Head, A Cup too much, Cock'd, Chipper, Loaded his Cart, Kill'd his Dog, Took his Drops, Wet both Eyes, Cock Ey'd, Got the Pole Evil, Got a brass Eye, Fox'd, Fuddled, Sore Footed, Well in for 't, Owes no Man a Farthing, Been to France, Flush'd, His Flag is out, Been at an Indian Feast, Groatable, Gold-headed, Glaiz'd, As Dizzy as a Goose, Had a Kick in the Guts, Been at Geneva, Got the Glanders, Jolly, Jagg'd, Going to Jerusalem, Jocular, Been to Jerico, Juicy, Seen the French King, Het his Kettle, He makes Indentures with his Leggs, Light, Limber, Merry, Moon-Ey'd, Muddled, Maudlin, Nimptopsical, Got the Night Mare, He's Oil'd, Smelt of an Onion, Oxycrocium, Pidgeon Ey'd, As good conditioned as a Puppy, Been among the Philistines, In his Prosperity, Eat a Pudding Bagg, Raddled, Lost his Rudder, Ragged, In the Sudds, As Drunk as David's Sow, Swampt, Staggerish, carries too much Sail, Stew'd, Soak'd, with all his Studding Sails out, Tongue-ty'd, Topsy Turvey, Tipsey, Got the Indian Vapours.

By comparison, modern depictions of the drunk are few and tame.

Only a man who loved words would have gone to this trouble, or enjoyed himself so much in the acquisition of so many of them. Franklin was one.

> The Phrases in this Dictionary are not (like most of our Terms of Art) borrow'd from Foreign Languages, neither are they collected from the Writings of the Learned in our own, but gather'd wholly from the modern Tavern-Conversation of Tiplers. I do not doubt but that there are many more in use; and I was even tempted to add a new one my self under the Letter B, to wit, Brutify'd: But upon Consideration, I fear'd being guilty of Injustice to the Brute Creation, if I represented Drunkenness as a beastly Vice, since, 'tis well-known, that the Brutes are in general a very sober sort of People.[5]

Though he indulged his whimsical impulses in these occasional pieces, Franklin hoped his words would have an impact on the real world. There is nowhere better to see this serious side in his early writings than in his May

1743 proposal for "Promoting Useful Knowledge among the British Planta-tions in America" that would lead directly to the founding of the American Philosophical Society (APS). The Society is still there, in a colonial brick structure on Fourth and Chestnut, across the side street from Independence Hall. Its collections are superb, its staff helpful, and its grounds redolent of Franklin's city.

In part, the proposal read like the soberer passages in "The Printer to the Reader": "The first Drudgery of Settling new Colonies, which confines the Attention of People to mere Necessaries, is now pretty well over; and there are many in every Province in Circumstances that set them at Ease, and af-ford Leisure to cultivate the finer Arts and improve the common Stock of Knowledge." This was an audacious claim for an inhabitant of a colony at the periphery of the British Empire, but Franklin's view was typical of the growing sense of autonomy of the provincial intelligentsia: "To such of these who are Men of Speculation, many Hints must from time to time arise, many Observations occur, which if well-examined, pursued and improved, might produce Discoveries to the Advantage of some or all of the British Planta-tions, or to the Benefit of Mankind in general."

The problem in Franklin's mind was not the readiness of American minds to contribute to the store of science and technology, but the vast distances separating thinkers and tinkerers in Boston from those in Charles Town. "But as from the Extent of the Country, such Persons are widely separated, and seldom can see and converse or be acquainted with each other, so that many useful Particulars remain uncommunicated, die with the Discoverers, and are lost to Mankind." The solution was communication through publications that circulated through the colonies. "It is, to remedy this Inconvenience for the future, proposed, THAT One Society be formed of Virtuosi or ingenious Men residing in the several Colonies, to be called *The American Philosophical Society*, who are to maintain a constant Correspondence."

Where should this intellectual institution reside? Of course, in Philadel-phia: "THAT Philadelphia being the City nearest the Centre of the Conti-nent-Colonies, communicating with all of them northward and southward by Post, and with all the Islands by Sea, and having the Advantage of a good growing Library, be the Centre of the Society." Franklin was the prototype of what would become an American icon—the city booster. After the novelist Upton Sinclair satirized the businessman George F. Babbitt, in his 1922 book of the same name, the "booster" became synonymous with conformity, social climbing, and materialism. But Franklin's boosterism was of a different sort. He sought to elevate, not flatten, the quest for knowledge.

Franklin was addicted to micromanagement. Order was one of the virtues he had imposed on himself and touted to others.

> THAT at *Philadelphia* there be always at least seven Members, viz. [namely] a Physician, a Botanist, a Mathematician, a Chemist, a Mechanician, a Geographer, and a general Natural Philosopher [scientist], besides a President, Treasurer and Secretary. THAT these Members meet once a Month, or oftener, at their own Expence, to communicate to each other their Observations, Experiments, &c. [etc.], to receive, read and consider such Letters, Communications, or Queries as shall be sent from distant Members; to direct the Dispersing of Copies of such Communications as are valuable to other distant Members in order to procure their Sentiments thereupon, &c.

There is neither frivolity nor satire in any of this proposal. It bespeaks a man of letters, science, and above all modern administrative sentiments proposing the creation of an institution at the center of all modern inquiry—a scientific society. Unlike the Royal Society in England, however, there was no appointment process, no government sponsorship, no limit on the number or the social status of members. It was voluntary, private, and open to all who wanted to contribute to it or benefit from it.

Franklin then proposed specifics, much as he had catalogued the objects donated to Whitefield's orphanage: "THAT the Subjects of the Correspondence be: All new-discovered Plants, Herbs, Trees, Roots, &c. their Virtues, Uses, &c. Methods of Propagating them, and making such as are useful but particular to some Plantations more general." Above all, the contributions must be useful. Franklin was never interested in metaphysics, no more than in the supernatural elements of revealed religion. Use, improvement, functionality—these were his lodestone. Useful knowledge had value.

> New Methods of Curing or Preventing Diseases. . . . New Mechanical Inventions for saving Labour; as Mills, Carriages, &c. and for Raising and Conveying of Water, Draining of Meadows, &c. All new Arts, Trades, Manufactures, &c. that may be proposed or thought of. Surveys, Maps and Charts of particular Parts of the Sea-coasts, or Inland Countries; Course and Junction of Rivers and great Roads, Situation of Lakes and Mountains, Nature

of the Soil and Productions, &c. New Methods of Improving the Breed of useful Animals; Introducing other Sorts from foreign Countries. New Improvements in Planting, Gardening, Clearing Land, &c.

But Franklin's conception of utility was not narrow- or present-minded. For him technology and science were one; hence, the American Philosophical Society was to promote "all philosophical Experiments that let Light into the Nature of Things, tend to increase the Power of Man over Matter, and multiply the Conveniencies or Pleasures of Life." Finally, Franklin could not resist one bit of whimsy. "Benjamin Franklin, the Writer of this Proposal, offers himself to serve the Society as their Secretary, 'till they shall be provided with one more capable."[6]

The modern manager is a bureaucrat. He or she promotes, arranges, publicizes, facilitates, raises funds, accounts for the expenditure of funds, and does it all in a public, orderly way. Franklin's proposal for the founding of the APS was a masterpiece of early modern bureaucratic thinking. The term bureaucrat was actually contemporary with his efforts, a name for the increasing number of French officials whose duties were administrative. With the growth of the French Empire, a parallel development to the English Empire (they had eyes on one another), the rise of officialdom was inevitable. In England, the number of full-time bureaucrats was keeping pace with the French, collecting taxes, keeping records, buying and cataloging ordnance—most of them to service the needs of the empire and the armed forces necessary for defending that empire.[7]

The spread of the Anglo-Spanish War from the Caribbean to the North American continent when France entered the war on Spain's side in 1744 spurred Franklin to a second highly serious proposal. In 1747 he urged the creation of a colonial militia to guard its frontiers. The Quaker majority in the assembly, opposed to war and all preparations for war, was obdurate about creating and arming a militia. Franklin recalled, "The laboured and long-continued Endeavours [to make] Provisions for the Security of the Province having proved abortive, I determined to try what might be done by a voluntary Association of the People. To promote this I first wrote and published a Pamphlet, entitled, PLAIN TRUTH, in which I stated our defenseless Situation in strong Lights, with the Necessity of Union and Discipline for our Defense, and promis'd to propose in a few Days an Association to be generally signed for that purpose."

Franklin did not stop with publication. By now he was no longer the shy teenager of Silence Dogood or the upstart printer of Busy-Body. He later recalled that "the Pamphlet had a sudden and surprising Effect. I was call'd upon for the Instrument of Association. And having settled the Draft of it with a few Friends, I appointed a Meeting of the Citizens in the large Building before-mentioned." Franklin was an able public speaker, though hardly the equal of Whitefield. "The House was pretty full. I had prepared a Number of printed Copies, and provided Pens and Ink dispers'd all over the Room. I harangu'd them a little on the Subject, read the Paper and explain'd it, and then distributed the Copies which were eagerly signed, not the least Objection being made."

Oddly, because everyone knew its author, his *Plain Truth* essay was signed "a Philadelphia Tradesman." Was it possible that Franklin, knowing the fate of others who directly assaulted the authority of the colonial legislature (censure and fines, even incarceration), opted for the pseudonym to prevent a contempt proceeding in the lower house? In 1773, for another breach of Parliamentary privilege (sending back to the colonies letters written by a royal governor of Massachusetts), Franklin faced a day-long public interrogation and humiliation. In any case, even those not at the meeting would have seen through the pseudonym immediately. For example, Charles Norris, a leading Quaker merchant and son of Isaac Norris, with whom Franklin had served as an agent of the government in negotiations with the western Pennsylvania Indians, obtained the pamphlet and wrote Franklin's name on the title page. In sum: always prudent, Franklin took precautions. But everyone knew who wrote it. Governor Thomas Penn, who had little use for Franklin or his politics, would have taken some action against the author, but he admitted that as Franklin was "the tribune of the people," nothing could be done.[8]

On some occasions, the serious side of Franklin's thinking adopted a style close to Whitefield's. The *Plain Truth* piece was such an occasion. It was a political sermon in its language and its portent, warning of a dire fate for those who did not see the light in time to save themselves. "War, at this time, rages over a great part of the known world; our newspapers are weekly filled with fresh accounts of the destruction it everywhere occasions." Franklin exaggerated: unlike the French and Indian War to come in 1754, King George's War was not a world war. But it was true that the Quaker majority in the Pennsylvania Assembly made no preparations for war, Quakers being pacifists. "Pennsylvania, indeed, situate in the centre of the colonies, has hitherto enjoyed profound repose; and though our nation is engaged in a bloody war, with two great and powerful kingdoms, yet, defended, in a great degree, from the French, on the one hand, by the northern provinces, and from the Spaniards, on the other, by the southern, at no small expense to each, our people have, till lately, slept securely in their habitations."[9]

Pennsylvania had seen warlike struggle on its frontier. The Iroquois, allies of the British in New York, had swept down on the Indians of Pennsylvania and the Ohio River Valley, with repercussions for settlers in the western portion of the colony. Treaties between the Penns and the Leni Lenape or Delaware Indians had prevented the kind of frontier warfare that roiled the Virginia, Carolina, and New York frontiers, but it was always touch and go for the negotiators, translators, and go-betweens keeping the peace in western Pennsylvania. And Franklin never really trusted the Indians. Visiting them during treaty negotiations, he recalled "their dark colored bodies, half naked, seen only by the gloomy light of the Bonfire" and their "terrible shriek and halloo," which "seemed to him formed a scene the most resembling our ideas of hell that could be imagined."[10]

What would happen if that hell were visited on the farms and villages of the colony? It had happened elsewhere during King George's War, particularly in the Lake Champlain region of New York. "The French and their Indians, boldly and with impunity, ravage the frontiers of New York and scalp the inhabitants." Indian warfare followed none of the rules of war that European professional forces had developed in the eighteenth century. There was no distinction between soldier and civilian, no sparing the weak or the wounded. Some captives the Indians took on their raids might be exchanged or ransomed, but others might be executed on the spot. (To be sure, the colonists were no respecters of Indian women and children, killing indiscriminately when they raided Indian villages.) As a result of the war of each against all on the frontier, "there is no British colony, excepting this, but has made some kind of provision for its defence; many of them have therefore never been attempted by an enemy; and others, that were attacked, have generally defended themselves with success."[11]

Philadelphia was a ripe target for Britain's enemies. "Our wealth, of late years much increased, is one strong temptation, our defenceless state another, to induce an enemy to attack us." In wartime, the polyglot population of the city made it even more vulnerable, "by means of the prisoners and flags of truce they have had among us; by spies which they almost everywhere maintain, and perhaps from traitors among ourselves; with the facility of get-ting pilots to conduct them." Colonists in the port cities of Boston, Newport, New York City, Philadelphia, Charles Town, and Savannah all lived in fear of a French or Spanish fleet sailing into the harbor, bombarding the piers and warehouses, sparking a slave rebellion, and generally doing mischief. When slaves rose up in rebellion in South Carolina in 1739 and in New York City in 1741, some among them expected or hoped for aid from Britain's imperial enemies. White authorities blamed the slave uprisings, in part, on Spanish agent provocateurs.

What about the countryside? Travel two days west or northwest from Phil-adelphia and one would enter Indian country. "We have, it is true, had a long peace with the Indians; but it is a long peace indeed, as well as a long lane, that has no ending. The French know the power and importance of the Six Nations, and spare no artifice, pains, or expense to gain them to their inter-est. By their priests they have converted many to their religion, and these have openly espoused their cause." Franklin's tolerance had limits. Roman Catholicism he viewed with horror. "Are there no priests among us, think you, that might, in the like case, give an enemy as good encouragement? It is well known, that we have numbers of the same religion with those, who of late encouraged the French to invade our mother country." The Six Nations were Iroquoian-speaking peoples of New York, allied to the English when it suited their needs, friendly to the French when diplomacy required it. With English settlers filtering into Iroquois lands in New York, their loyalty to Britain might be tested. Indian diplomacy "never allowed the colonizers to be certain that Iroquois arms would not be turned against them."[12]

The other Indians of the Upper Ohio Valley had been allied with the French in earlier frontier wars. The French were now courting them with presents, favorable trade terms, and warnings about the land hunger of the English-speaking settlers. "The rest [of the Indians] appear irresolute what part to take; no persuasions, though enforced with costly presents, having yet been able to engage them generally on our side." In fact, the Pennsylva-nia traders, led by the notoriously corrupt George Croghan, were bilking the Indians and lying about it. In 1748, the same Croghan would be carrying on the negotiations.[13]

How could Franklin persuade the assembly and the people at large of the threat? In the end, he adopted a political soteriology as dire as any that Whitefield advanced, and like Whitefield's, Franklin's essay on how to be saved was rooted in a Bible story.

> That our enemies may have spies abroad, and some even in these colonies, will not be made much doubt of, when it is considered, that such has been the practice of all nations in all ages, whenever they were engaged, or intended to engage, in war. Of this we have an early example in the Book of Judges (too pertinent to our case, and therefore I must beg leave a little to enlarge upon it), where we are told . . . that the children of Dan sent of their family five men from their coasts to spy out the land, and search it, saying, Go, search the land.

Franklin had an eye for a riveting tale, a tale of nations at war and spies abroad, a tale of the wages of unpreparedness. Like those sinners who slumbered in false hope whom Whitefield hoped to rouse to their danger, Franklin exhorted the people of Pennsylvania: "These Danites . . . came to Laish, and saw the people that were therein, how they dwelt Careless. . . . They thought themselves secure, no doubt; and as they never had been disturbed, vainly imagined they never should. It is not unlikely, that some might see the danger they were exposed to by living in that careless manner; but that, if these publicly expressed their apprehensions, the rest reproached them as timorous persons, wanting courage or confidence in their gods, who (they might say) had hitherto protected them."

So the people of Laish refused to see the danger even when some among them (a Franklin, perhaps) sounded the alarm. "The spies returned, and said to their countrymen, Arise, that we may go up against them; for we have seen the land, and behold it is very good. And are ye still? Be not slothful to go. . . . When ye go, ye shall come to a people secure, (that is, a people that apprehend no danger, and therefore have made no provision against it)," to which Franklin commented, in an aside, "great encouragement this!" For apparently Laish was "a large land, and a place where there is no want of any thing" (a clear reference to Pennsylvania).

The result was predictable, at least as far as the Danites were concerned. "Accordingly we find in the following verses, that six hundred men only, appointed with weapons of war, undertook the conquest of this large land." Just in case the analogy was still unclear to his readers (after all, they had up to that point ignored the threat), Franklin brought the Catholic Church into the story: "the idolatrous priest, with his graven image, and his ephod, and his teraphim, and his molten image (plenty of superstitious trinkets) joined with them, and, no doubt, gave them all the intelligence and assistance in his power." The Danites "smote them with the edge of the sword, and burnt the city with Fire; and there was no Deliverer, because it was far from Zidon.— Not so far from Zidon, however, as Pennsylvania is from Britain."

The lesson was crystal clear, or should have been, to just about every reader by now. "As the Scriptures are given for our reproof, instruction, and warning, may we make a due use of this example, before it be too late!" Scripture, exegesis, application: arm against the foe—not Satan this time, but the French and their Indian allies. The analogy seemed compelling to Franklin, the danger not of damnation in the next world but of safety in this one.

> Perhaps some in the city, towns, and plantations near the river, may say to themselves, "An Indian war on the frontiers will not affect us; the enemy will never come near our habitations; let those concerned take care of themselves." And others who live in the country, when they are told of the danger the city is in from attempts by sea, may say, "What is that to us? The enemy will be satisfied with the plunder of the town, and never think it worth his while to visit our plantations; let the town take care of itself."
>
> . . . Are these the sentiments of true Pennsylvanians, of fellow-countrymen, or even of men that have common sense or goodness?

Franklin pushed all the same emotional buttons that Whitefield pushed; indeed, Franklin was reminded (as if a child reared in a Puritan New England family needed reminding) by Whitefield about the terrors in the night that the unconverted surely would feel when judgment was upon them. "And should the enemy, through our supineness and neglect to provide for the defence both of our trade and country, be encouraged to attempt this city, and, after plundering us of our goods, either burn it, or put it to ransom, how great would that loss be! besides the confusion, terror, and distress, so many hundreds of families would be involved in!" Like Whitefield, Franklin personalized the Bible story, making it his and his readers'.

He was moved by it, as Whitefield was by his scriptural texts. "The thought of this latter circumstance so much affects me, that I cannot forbear expatiating somewhat more upon it. . . . Hence, on the first alarm, terror will spread over all; and, as no man can with certainty depend that another will stand by him, beyond doubt very many will seek safety by a speedy flight. . . . All will run into confusion, amidst cries and lamentations, and the hurry and disorder of departers, carrying away their effects . . . what must be your condition, if suddenly surprised, without previous alarm, perhaps in the night!"

Franklin had no doubt read the most famous Puritan poem of the previous century, Michael Wigglesworth's "Day of Doom" (1666):

> Still was the night, serene and bright,
> > when all men sleeping lay;
> Calm was the season, and carnal reason
> > thought so 'twould last for aye.
> Soul, take thine ease, let sorrow cease,
> > much good thou hast in store:
> This was their song, their cups among,
> > the evening before.
>
> .   .   .   .
>
> For at midnight break forth a light,
> > which turn'd the night to day,
> And speedily an hideous cry
> > did all the world dismay.
> Sinners awake, their hearts do ache,
> > trembling their loins surpriseth;
> Amaz'd with fear, by what they hear,
> > each one of them ariseth.

It was the same terror that Whitefield embedded in his sermons.

Franklin's words were spoken to an assembly of his fellow Philadelphia citizens, a secular meeting called to deal with this-world dangers of Indian war. *Plain Truth* was a democratic document in the sense that it was not just directed by Franklin to those in power. Franklin hoped to influence the Assembly by appealing to its constituents, directly to the voters. By holding an open meeting, and by circulating his *Plain Truth* among them, he relied on the sense of self-preservation of the people to change government policy. The tie between the policy in his actions and the policy in the publication was that an aroused citizenry could change the policy of the government. In a time when deference in politics to one's "betters" was the rule and "democracy" was an epithet applied to an unruly mob, this was a bold stroke. It was, as it would prove, also prophetic.[14]

Franklin the private man of business was becoming Franklin the man of public affairs. He relished this public role, fabricating new masks to wear as he performed it. He was becoming the Franklin that history knows, the Enlightenment Man.

A Great Awakening, the Enlightenment, and the Crisis of Provincialism

five

WHEN NEXT FRANKLIN AND WHITEFIELD MET, the British North American colonies appeared to have fulfilled all the expectations of the home country. Britain was growing wealthy and powerful on the exports from its American colonies. The colonies themselves grew more British as imports of books, furniture, fabrics, and all the "baubles of Britain" filtered down from the elite to the commoner. The "guardians" of tradition and orthodoxy in both metropolitan center and colonial periphery were finding it harder and harder to manage the expansive impulses of American culture. All manner of new and disturbing ideas were abroad, stirring minds and hearts, making people think a lot harder about what they believed, and why. Two of these provocative ideas were "Awakening," the catchword for a general reformation and revival of religion, and Enlightenment, an ideal of a progressive and naturalist reformation of intellectual life based on science. Whitefield and Franklin found themselves leading spokesmen and exemplars of the two notions, respectively.[1]

Whitefield's third tour in America, in 1740, brought his charismatic and unique style of preaching to a place where itinerancy and revivalism were already beginning to blossom. The New England and Middle Atlantic colonies were experiencing what Whitefield would call "a great and general awakening." Its churches were hives of contention. Some of the New Light fomenters of that contention, particularly Pennsylvania Presbyterian minister Gilbert Tennent, would meet and greet Whitefield after his first visit to Philadelphia. Whitefield and Tennent hit it off, becoming something of a mutual admiration society, particularly after Whitefield's triumph in New England in the fall of 1740. Whitefield would latter praise Tennent's style of preaching—the Pennsylvanian was a veritable "son of thunder" on the pulpit. Whether Tennent, who already had announced that some members of the ministerial fellowship were unregenerate and should not be allowed to preach, inducted Whitefield into that controversy and spurred Whitefield's already practiced "immoderation" when it came to doctrinal disputes, or Whitefield himself saw which side of the controversy he preferred and eagerly became a major figure in it, remains a question that defies a certain answer. But it is certain that "New Englanders took to Whitefield's message because it was already their message, the message they had been born and bred to hear, newly condensed in content and tricked out in images unusual in sermons but all too real to a people whose much reprinted classic was *The Day of Doom*." Indeed, it was part of New England's Puritan religious heritage that when the millennium came, New England would be home to the saving remnant.[2]

Historians debate when (and sometimes if) the Great Awakening began. When did the dour and backward-looking jeremiad become the enthusiasm of evangelical preaching? So-called covenant renewals had brought congregations together in the closing decades of the seventeenth century. But the preaching in these was old style, with the minister reading long sermons from the pulpit. In the early decades of the eighteenth century a new wave of pietistic enthusiasm swept through Protestant Western Europe and the British Isles. Some of these waves lapped upon the New England shore in the 1730s. Young people in particular proved ready for regeneration. At Taunton, Massachusetts, Minister Samuel Danforth reported that many were coming to him asking how to be saved. The youths in Gloucester, Massachusetts, were similarly troubled. But the most "remarkable" of the remarkable occurrences of renewed enthusiasm came in Northampton, where young Jonathan Edwards ministered in his late grandfather Solomon Stoddard's church. Edwards's report of the revival in the town, *A Faithful Narrative of the Surpris-*

ing Work of God (1737), spread like a brush fire throughout New England. "The young people showed more of a disposition to harken to counsel, and by degrees left off their frolicking and . . . there were more that manifested a religious concern than there used to be. . . . The young people declared themselves convinced by what they heard from the pulpit." Edwards reached out to the young people, and they responded to him. The shift from older authority to youthful yearning, from the formalities of pulpit preaching to impromptu settings for study, and the increasingly personalized relationship between congregant and minister would all become marks of the Awakening. In the course of this revival, men and women assembled to hear preachers in meetinghouses, on city streets, and in country fields, "as if they were hearing the gospel and seeing the world for the first time." Sometimes the gathering listened in absolute silence. More often, the minister's words were soon accompanied by the groans of the penitent sinner and the outcries of the awakened seeker.[3]

Whitefield arrived in Newport, Rhode Island, in the fall of 1740, traveled to Boston, out to Cambridge and Harvard College, on to Northampton, where Edwards met him, then down to New Haven and Yale College. Nothing, it seemed to the young listeners who flocked to hear him, would ever be the same. Like all whirlwinds, he left in his wake debris—but his converts among the young were as hot to defend him as his detractors were to criticize. Those who were disposed to hear in his words a consistent Calvinism, like Samuel Hopkins (who heard Whitefield preach at Yale that fall), believed that Whitefield taught "the doctrine of election that God . . . has chosen a certain number of mankind to be redeemed, fixing on every particular person whom he will save, and giving up the rest to final impenitence and endless destruction." But others, like John William Fletcher, a Methodist and critic of Whitefield, heard something quite different: "Whitefield understood far better how to offer up a warm prayer, and preach a pathetic sermon, than how to follow error into her lurking holes." In short, Whitefield lacked "consistency" between his doctrinal views and the message he offered his listeners (though even Hopkins believed that Christ died for all sinners, not just those to whom particular election applied). For some of Whitefield's audience, if not for those who shared his doctrinaire side, the injunction to be born again in Christ was something of a shortcut to salvation entirely different from the long and agonizing preparation New England Puritanism imposed on its church members.[4]

It was Whitefield's "immoderate" participation in the quarrel between the New Lights and their opponents that paved the way for his jubilant welcome among the revivalists rather than his strict Calvinism. Whitefield's presence was by the end of 1740 central to the Great Awakening's profound transformation of American styles of worship. These too expanded the Calvinism of his ministerial allies beyond its doctrinal limitations. First and perhaps foremost, he preached to the unconverted, not to the full members of the visible churches. He was in this sense unconcerned with churching, or at least relatively indifferent to the divisions within congregations his message might exacerbate, perhaps because he had never had his own pulpit in England.

His impromptu congregations included women, slaves, laborers—in fact, anyone who wanted to come and listen could. The gathering was thus entirely (well almost entirely) voluntary. Even the gathering of self-selected saints in the first Puritan great migration to New England was not entirely voluntary. Women came because their fathers or husbands brought them. Servants and children came because their masters and parents commanded them. Not so the crowds around Whitefield. They came because they wanted to, and no one could stop them (though some tried).[5]

The effect of his preaching was similarly novel. Dismissed as mere "enthusiasm" or, worse, the work of a charlatan, the emotional response of his listeners was genuine. While emotion expressed during the meeting was an important part of Christian worship, it did not have a central role until Whitefield began his crusade. When he preached, he freely expressed his own emotion, asking God to be his helper, admitting his weakness, and crying genuine tears. His auditory mimicked these emotions, also genuinely. Thus, the revival was a cathartic experience as well as a theological one. This kind of intense, personal experience had a special resonance in the Puritan churches of New England, for full membership in them required a candidate to narrate to the congregants his or her spiritual journey. As Thomas Prince, a New Light ally of Whitefield, wrote shortly after Whitefield departed in 1740, "within three months, were three score joined to our communicants, the greater part of whom gave a more exact account of the work of the spirit of God upon their souls."[6]

While his theology was in many ways a throwback to earlier, stricter formulations of Calvinism, Whitefield's rhetorical message that rebirth alone made one a true Christian opened doors to salvation that a more traditional

Calvinism lacked. He differed with the Methodists on how many might have been elect (Wesley's insistence that Christ had died for everyone) and with the Church of England over its apparent indifference to predestination (although the Church never formally renounced its Calvinist origins, few of its clergy were strict Calvinists), but the impact of his preaching was to open the door to anyone who wanted to enter. Though he insisted that no one could will their way into Heaven—indeed, the exact opposite, that men and women were helpless to change God's mind—the wholeness of the revival experience carried a different message. Rebirth was the first step that a person could take on the road to salvation.

Whitefield's rejection of the rational or reasonable doctrines of regeneration held by most ministers of the Church of England became more virulent in the early 1740s. This put him on a collision course with Anglican commissary Alexander Garden of Charles Town and New England Old Lights like Charles Chauncy. Garden had earlier welcomed Whitefield, but by 1740 he had become an implacable critic. Garden particularly found offensive Whitefield's implication that "if to these doctrines you demur and object, that you do not apprehend them sufficiently ground in the holy Scriptures, or ever taught by the Catholick Church of Christ [i.e., here the Church of England] in any Age—the reason is, they'll answer you, because you are an unregenerate person—you have not the spirit of God discerned, but you see and judge of spiritual things, only by the eyes of your carnal and corrupt reason." The two men never reconciled. Chauncy had remained more or less above the fray until 1742, when he jumped in with both feet. The revivalists "were conceited of their gifts, and too generally disposed to make an ostentatious show of them." Their "vain glorious temper" had led them to confuse "disturbance" with conversion. Whitefield was in the vanguard of this crew.[7]

Did Whitefield know that his words convinced people that they could take some part in being saved in spite of his Calvinistic doctrines? He had a great many chances to pray, sing, and converse with sincere followers. His journals report that he was able to help many of these penitents find the way to be good Christians. What use was that path (and why did so many seek personal audiences with him) if he told them that there was no hope for them? Instead, he told them that God loved them, and that gave them hope. In this fashion, Whitefield was a kind of psychologist. For the ills of the spirit he preached positive thinking. He convinced thousands that they were not helpless. "The sun shown upon us; and I trust the sun of righteousness arose on some with healing on its wings." Modern clinical psychology has the same aim—to restore the patient's self-image and hope for the future.[8]

The size of his audiences, his ability to be heard, the way in which he used publicity to gain a mass following—all of these were phenomena in themselves that kept the Awakening alive. And when these seemed to pall, Whitefield himself stirred the ashes of enthusiasm to rekindle the revival. His 1744–1748 tour of the colonies was dogged by controversy. For deeply imbedded in the evangelical vision of the world then, as now, was a sense of the differences between true converts and those whose worship was mere form—what the most radical of the first Puritans in England would have called legal Christians, going through the motions by going to church, reciting the words without feeling the Holy Spirit. The saved knew who they were through the testimony of the Word, spoken aloud, in visible gatherings; they were assured of their salvation through their striving for piety and their rejection of the temptations of heresy and materialism. Their ministers were a fellowship of "burning and shining lights," according to Whitefield. Those who did not see the light were "dead drones." Their critics regarded the evangelicals as misguided at best and hysterical at worst. Perceptions of otherness again divided communities and congregations into New and Old Lights (among Congregationalists), New and Old Sides (among Presbyterians), Anglicans from Methodists, and Baptists from just about every other Protestant sect.[9]

Contemporary critics may have had a point when they wailed about the enthusiasm of the New Lights. Some of the more extreme manifestations of the revival involved James Davenport. Davenport was as immediately touched by Christ as Whitefield, or so he claimed. Davenport's revival meetings included bursts of song and shouts of joy. He and his followers marched through the streets of Boston in the summer of 1742 singing hymns at the top of their voices. Gesture and tonality were essential parts of preaching. Davenport's "song," published in 1742, implied that his intimate connection to Christ empowered his preaching. "I see thy face, I hear thy voice," he sang, "such hidden manna . . . as worldlings do not know, eye hath not seen, near ear hath heard." When he visited New London in March 1743, he enjoined his followers to throw their gaudy clothes and wigs into a bond fire, along with ungodly books. According to one less than complimentary eyewitness account, he shouted over the blazing fire, "thus the souls of the authors of those books, those of them that are dead, are roasting in the flames of hell." When, the next night, the bonfire was relit, Davenport called for an end to the idolatry of things—fancy clothing in particular. Raiment "worn for ornament" must be purified in the cleansing flames. Davenport contributed his britches, saying "go you with the rest." Skeptics complained that Davenport's extemporaneous preaching was "scarcely worth the hearing. The Praying was without form or comeliness. It was difficult to distinguish between his praying and preaching for it was all mere confused medley." It was the same complaint that Whitefield's critics had been making for years about his preaching.[10]

The fever of the Awakening could not exist for long, and signs of its decline were everywhere apparent by the middle of the 1740s. Even Davenport had recanted some of the excesses of his ministry and adopted a learned, restrained style. In July 1744, as the revival movement ebbed, he published a short retraction and concession that his zeal was "misguided" and he should not have condemned his brethren or urged their congregations to denounce their pastors. But now, his passion had cooled, and with it, his mistaken "impressions" seemed to him an erroneous method. He should, he confessed, have stuck to the "analogy of scripture."[11]

Whitefield never retracted or retreated from his view of the Awakening, but with its demise, he conceded that "a chill came over the church's work" in New England, and those who had welcomed him in the first days of 1740 were by 1744 apprehensive that his preaching would foment "separations" in congregations lately returned to harmony. He was unmoved by their appeal, replying, "How can a dead man beget a living child." He refused to admit the obvious, gained permission from three churches to guest on their pulpits, and stirred the separations that Old Light preachers condemned. He remained a divisive figure, and after the early 1740s, his periodic return to New England did not "generate widespread interest." In the meantime, the "hostility" that trailed his ceaseless traveling caught up to him in England. The bishops of the Church of England had not forgotten his cavalier disregard of their admonitions. While they hurled thunderbolts from above, a growing number of hecklers interrupted his meetings. On more than one occasion, he found himself facing an angry mob. His ministry had always been strongest among the middle classes, and they did not desert him, but he was no longer the boy wonder.[12]

One of the most common complaints against his mission was that he had converted contributions to the Georgia orphan house to his own use. In 1746, once again in Savannah, he felt obligated to report to Franklin, for publication in the *Gazette*, Bethesda's financial situation.

> As it is a Minister's Duty to provide Things honest in the Sight of all Men, I thought it my Duty, when lately at Georgia, to have the Whole Orphan House Accounts audited, from the Beginning of that Institution to January last; the same I intend to do yearly for the Future: An Abstract of the whole, with the particular Affidavits, and common Seal of Savannah affixed to it, I have sent you with this; be pleased to publish it in your weekly Paper. My Friends thought this was the most satisfactory Way of proceeding. To print every particular Article, with the proper Voucher, would make a Folio, and put me to a greater Expence than my present Arrears will permit me to be at.

He had turned to another form of witnesses, civil rather than religious, an accounting of monies rather than souls, to acquit himself.

Whitefield reported cash "received from the 15th December, 1738, to the 1st January, 1745–6, by publick Collections, private Benefactions, and annual Subscriptions" as 5,511 pounds sterling, and expenses for "Buildings, Cultivation of Lands, Infirmary, Provisions, Wearing Apparel, and other incident expences," as the very same 4,982 pounds sterling. The auditors, including James Habersham, swore that "the Reverend Mr. Whitefield further declareth, that he hath not converted or applied any Part thereof to his own private Use and Property, neither hath charged the said House with any of his travelling, or any other private Expences whatsoever." To corroborate Habersham, who was Whitefield's convert as well as his associate in the venture, the audit was attested to by five Savannah witnesses,

> who being duly sworn, say, That they have carefully and strictly examined all and singular the Accounts relating to the Orphan House, in Georgia, contained in Forty One Pages, in a Book entitled, Receipts and Disbursements for the Orphan House in Georgia, with the original Bills, Receipts, and other Vouchers . . . and that it doth not appear that the Reverend Mr. Whitefield hath converted any Part thereof to his own private Use and Property, or charged the said House with any of his travelling, or other private Expences; but, on the contrary, hath contributed to the said House many valuable Benefactions.[13]

Whitefield had made concessions to his secular critics, for example, supporting the institution of slavery in the South. Baptist and Methodist ministries in early America openly condemned slavery, but both of these sects' luminaries in the South retreated from their initial stance, their ministers by the 1830s becoming the most consistent defenders of slavery. Whitefield himself accepted gifts of slaves, and "his new popularity among the southern gentry was, in fact, directly proportional to his endorsement of slavery."[14]

Through his thirties and forties, he had become an institution, still periodically touring the sites of old triumphs, still writing of the revival as if it were fresh. His journals for these years reproduced the language of the first journals; indeed, the needle was stuck in the same groove. He spent more and more of his time in America, a permanent itinerant, welcomed everywhere, but at home in no one place, not even Bethesda, though he tarried there when he could.

Then something happened. Perhaps it began when William Seward was gone—killed by an angry mob in Wiltshire, England. Whitefield himself received death threats. Or perhaps it began in Boston, as the revival cooled. In 1744, during his third visit to New England, he found himself the target of conservative preachers and had to back away from the criticism of an unconverted ministry that had been a centerpiece of his preaching since 1740. One of his Boston allies, Minister Benjamin Colman, judged, "Mr. Whitefield seems to grow more and more sensible of the evil spirit of animosity among us and the worth of our ministers."

Perhaps it was a sense that he had written himself out—all of the New Lights had—and needed to think more carefully about what he allowed of his thinking to appear. "The number of Whitefield's publications coming off colonial presses declined sharply." In 1756 he supervised a new edition of his writings and excised the more provocative and censorious passages from his journals. He was still the master of the revelatory journal piece, the self-referring sermon, still the center of the whirlwind of words, but he was aware that he could no longer claim youthful indiscretion or passion as an excuse to insult his clerical brethren. Whitefield was growing older, and a more mature sensibility dictated greater sobriety.[15]

Or perhaps the reason for his changing personality was the fact that he acquired a patroness, Lady Selina Hastings, the widowed Countess of Huntingdon, and from 1747 until he died he was her personal chaplain, though they were often an ocean apart. In 1739 she had become an adherent of the

Gate to Bethesda Orphanage, Savannah, Georgia, in the eighteenth or nineteenth century. The orphanage was the joy and travail of Whitefield's life. Its children were his surrogate family. Georgia Historical Society.

Methodist John Wesley, and then of Whitefield. There are nearly a hundred letters from him to her, more than to any other correspondent, including Franklin, the Wesley brothers, and Habersham. Though couched in terms of their common religious faith, they hint at a deeply shared personal affection. Her patronage opened doors for him to preach to the upper classes. At her home on Park Street in the London borough of Westminster, they received ministers of state, including the prime minister, Robert Walpole. "On this first visit to Scotland [Whitefield] was most hospitably received by many persons of rank, who behaved towards him with great politeness and attention: and this attention was considerably increased in every subsequent visit, after he became chaplain to the Countess of Huntingdon; her Ladyship being, as we have already shown, well known to many of the Scotch nobility, among whom she had a very extensive acquaintance."[16]

Apprised of Whitefield's new friends among the rich and powerful, Franklin was slyly reproving. He wrote to Whitefield on July 6, 1749, a letter closer in tone to one of his essays than to confidences shared with a close friend. But then, Franklin was simply being Franklin:

> I am glad to hear that you have frequent opportunities of preaching among the great. If you can gain them to a good and exemplary life, wonderful changes will follow in the manners of the lower ranks; for . . . On this principle Confucius, the famous eastern reformer, proceeded. When he saw his country sunk in vice, and wickedness of all kinds triumphant, he applied himself first to the grandees; and having by his doctrine won them to the cause of virtue, the commons followed in multitudes. . . . Our more western reformations began with the ignorant mob; and when numbers of them were gained, interest and party-views drew in the wise and great.[17]

In all, Whitefield would make seven trips to the colonies, interspersed with tours of Scotland and stops in Bermuda and Gibraltar. The revival had by the 1750s spread to backwoods Virginia, as Baptist and Methodist ministers reached out to settlers and slaves. No longer required to carry the word everywhere in person, Whitefield became a fundraiser for a variety of colonial institutions. He joined with Franklin in raising funds for schools in Philadelphia, as well as for a Presbyterian college in New Jersey (later Princeton College) and Eleazer Wheelock's school for Indians in Hanover, New Hampshire (which would later evolve into Dartmouth College). None of these were affiliated with his own Church of England, but then, his quarrels with its leaders had become almost a defining feature of his ministry.

Franklin's star had not risen as fast as Whitefield's in the first years of the 1740s, but by the end of the decade he was the better-known figure throughout the empire. Franklin sought to change the world around him rather than help others find their way to the next world. Indeed, it is hard to find a more secular thinker in the 1740s and 1750s than Franklin. In this, along with his faith in science, he was an avatar of the Enlightenment.

Peter Gay, the foremost modern student of the Enlightenment, describes it thus: "In the century of the Enlightenment, educated Europeans awoke to a new sense of life. They experienced an expansive power over nature and themselves: the pitiless cycles of epidemics, famines, risky life and early death, devastating war and uneasy peace . . . seemed to be yielding to the application of critical intelligence." Innovation, improvement, and science

George Whitefield, ca. 1768. Near the end of his days, Whitefield was still passion-ate in his calling. Portrait by John Greenwood, published by Carington Bowles, after Nathaniel Hone mezzotint.

all led the way to a better life and a better world. It was an optimistic move-ment, rooted in the belief in human capacity and intellectual progress. Such a "faith in progress" was perhaps naive and certainly too eagerly grasped at, but insofar as a faith in reason and improvement was tied to actual rise in standards of living, some progress was real enough. And "if America was the embodiment and natural home of the Enlightenment, according to Europe's advanced thinkers, then the American who best personified the Enlighten-ment ideal was Benjamin Franklin."[18]

The Enlightenment expressed itself in three major areas of intellectual endeavor: literature, scientific experiment, and political and economic reform. In literature, the Enlightenment explored more naturalistic forms of depiction, including the invention of the novel. In fact, it was the Enlightenment's insistence on the strict separation of fact from fiction that allowed the novel to blossom as a literary type. But historical novels, adventure novels set in far-off places, and novels of manners and morals all described secular events in everyday language. This explosion of fictional literature would not have been possible without a corresponding expansion of readership. Literacy became the norm rather than the exception for Enlightenment Europe and the British colonies.[19]

Alongside fiction, nonfiction genres evolved. Historical writing came into its own, with a canon stressing facts rather than fables and source checking. The learned essay on manners and morals, politics, and other popular topics was the mainstay of the new "magazine" or journal of opinion. Political tracts, freed from the constraints of earlier censorship, explored current and past regimes. Although an author could still be prosecuted for criticism of the current ruling family, disquisitions on political philosophy were relatively safe publications. For this reason, John Locke finally allowed his *Two Treatises on Government* to be published in 1690 (although he still kept his authorship secret). There followed essays such as Montesquieu's *Spirit of the Laws* and the works of Voltaire in France. Another of the new nonfiction publications was the encyclopedia. Ephraim Chambers's *Cyclopaedia* (1728) was the first. Finally, the newspaper, with its "news" and its opinion pieces, announced the triumph of fact over fancy and reportage over rumor and gossip. With "the lapse of the licensing act in 1695," newspapers were freed from the heavy hand of prior censorship. Newspaper publication and readership blossomed virtually overnight in enlightenment Britain and America.[20]

Franklin was a promoter and a beneficiary of both the new literature and the widened readership. He had learned to write essays by copying from one of the most popular of the early magazines, the *Spectator*. His own essays had a wide readership because he wanted to reach as many readers as he could. He had made his fortune with newspapers, his almanac, and other printing projects, including the arrangement with Whitefield. Franklin retired from active business in 1748, though he kept a hand in—making over six hundred pounds sterling a year as a non-managing partner from his print and publication operations. Add to this the official business that came to the newspaper (he had a contract to print official Pennsylvania legal documents), his loan office business, his land speculation, and other commercial deals, and the total was comparable to what other wealthy Americans (for example, John Hancock and George Washington) would make from their business operations. He was now by his own lights and by all measures but blood a "gentleman" of leisure, with a coat of arms to go along with it—all because of Enlightenment literary tastes.[21]

But leisure was hardly his goal. In the very years surrounding his retirement from business, he found himself engaging in the second of the Enlightenment's themes—scientific experiment. Indeed, it has been argued that the Enlightenment began with Isaac Newton's scientific discoveries. Universal natural laws such as gravity were observable, testable, replicable, and above all needed no divine intervention to work. Newton's world was a marvelous clock whose movements were regular and predictable if one knew the science. Natural law, not God's Word, ruled the world (though Newton was a religious mystic himself and an alchemist) in the Enlightenment.[22]

Franklin had no formal training in the sciences. His interests were eclectic. He never sought a patent for any of his inventions. Instead, he had "secured leisure during the rest of my life, for philosophical studies [i.e., natural philosophy, another word for science] and amusements." Among these were "my electrical experiments."[23]

Franklin became interested in electricity in much the same way as he joined forces with Whitefield, out of curiosity. He was not a modern scientist, studying for years in universities, doing postdoctoral research in a laboratory or for a firm. His was a science of the senses, of demonstrable relationships in real time among natural phenomena (heat, light, cold) that could be measured with some precision. After seeing Archibald Spencer's demonstrations of static electricity in Boston in 1743, Franklin arranged with the Scottish scientist to publicize his appearances that year in Philadelphia. The practical side of the display interested Franklin himself—particularly the use of glass jars to contain what appeared to be electrical energy. As he wrote to Peter Collinson, an English fellow amateur scientist, in 1747,

> Your kind present of an electric tube, with directions for using it, has put
> several of us on making electrical experiments, in which we have observed
> some particular phaenomena that we look upon to be new. I shall, there-
> fore communicate them to you in my next, though possibly they may not be

new to you, as among the numbers daily employed in those experiments on your side the water, 'tis probable some one or other has hit on the same observations. For my own part, I never was before engaged in any study that so totally engrossed my attention and my time as this has lately done; for what with making experiments when I can be alone, and repeating them to my Friends and Acquaintances, who, from the novelty of the thing, come continually in crowds to see them, I have, during some months past, had little leisure for any thing else.[24]

Franklin knew that the great tradition of British science, the science of biologist Robert Hooke, chemist Robert Boyle, and Newton, was experimentation. They were members of the Royal Society of London chartered in 1662. Indeed, the face of the society was public demonstration of phenomena. From its origins, the members of the society believed that "our experiments will bring to our public duties and actions" due applause and support. The reports of the Royal Society of London had been required reading for Franklin's circle. He had ordered copies for the library company and the Philosophical Society. He went to school, so to speak, on Newton's *Opticks*, in large measure because of its detailed descriptions of physical experimentation. The revelation of the secrets of the hidden world of light was for him the very essence of science.[25]

In other words, Franklin's science was mechanical rather than theoretical. His fascination with the apparatus was as much a motive for his experiments as his desire to understand the underlying physics of electricity. Thus, the bulk of his epistolary reports to Collinson described the apparatus, including in somewhat gruesome detail how he used human subjects (he charged them with static electricity, had them stand on different materials, and then had them shock one another in one of his experiments). "But I shall never have done if I tell you all my conjectures, thoughts, and imaginations on the nature and operations of this electrical fluid and relate the variety of the little experiments we have tried."[26]

Collinson was not only a willing and helpful correspondent; he was a successful businessman (in the cloth trade), an amateur botanist, a Quaker, a philanthropist (he helped establish a hospital for foundlings), and a sponsor of the American Philosophical Society and the Free Library Company (supplying books, among other aids, to the Philadelphians)—in short, someone like Franklin. Collinson's network of friends included Philadelphia botanist

John Bartram, New York's Cadwallader Colden (the first anthropologist in America), and such European luminaries as Carl Linnaeus. All of these men shared a curiosity about the laws of nature in Newton's universe.

At first Franklin worked on the periphery of this community of scientists. But he knew it was held together by the written word. Not only did the scientists correspond with one another, but they wrote and published their findings. Once again the way to fame lay through the printed word—in this case Collinson's publication of Franklin's *Experiments and Observations on Electricity, Made at Philadelphia in America* (1751). The little book (really not much more than Franklin's reports to Collinson, some eighty-six pages in print) "went through five English editions, three in French, one in Italian, and one in German."[27]

Franklin's experiments with electricity, published and read around the Enlightenment world, made him one of the most famous scientists of his era. Electricity was one of the great puzzles of Enlightenment science when Franklin took it up. In France, experimenters delighted the king with electrical demonstrations of various kinds but had no idea how they were linked. Older ideas of electricity associated it with fire. Franklin's experiments proved otherwise. From his experiments he concluded that electricity was a kind of fluid or current that each object living and inorganic possessed. "These appearances we attempt to account for thus: We suppose, as aforesaid, that electrical fire is a common element, of which every one . . . has his equal share." This insight alone should have gained Franklin immortality in physics. Electricity is the result of the movement of electrons, and every material body is composed in part of electrons. Franklin understood how opposing charges in this fluid attracted. His notion of electricity as fluid allowed for measurement of the flow of charges and further experimentation.[28]

If electricity was a kind of current, or flow of power, then it might be harnessed to use in more than popular demonstrations. But how to prove such a theory? The key to him and others working on the same problems was to visualize what they were theorizing. What if lightning, the most vivid example of electricity, could be the object of such a test? In 1752, Franklin attached a key to a kite and flew it in an electrical storm. Lightning attracted to the kite by the key was conducted through it to the kite. So it was a current. Franklin would add to the lexicon of science the same terms that are used today to describe the gross phenomenon of electricity—for example, "battery, conductor, condenser, charge, discharge, uncharged, negative, minus, plus, electric shock, and electrician."[29]

Franklin's further experiments on the storing of charges in different materials (capacitors) laid the groundwork for most early electronic devices, including the battery and the lightning rod. Lightning rods were soon as common in better European homes and commercial buildings as chimneys. Franklin received the Copley Medal of the Royal Society in 1753, as well as honorary degrees from Harvard and Yale Colleges. (He had never admitted that Silence Dogood's satirical raillery against Harvard came from his hand.)[30]

The mechanical bent of Franklin's experiments in electricity also showed itself in a variety of inventions. Chief among these was the so-called Pennsylvania Fireplace or Franklin Stove. Tinkering with devices was an old European habit, going back to the ancient Greeks and Romans. The stove was an improvement on the fireplace, throwing its heat all around the room, allowing for cooking on the top without the ashes of the fireplace singeing the bread and burning the meat. The sooty chimney above the fireplace, a fire waiting to happen, was replaced in the stove by the easily disassembled and cleaned flue and piping. There were a number of such stoves in Philadelphia area homes and shops when Franklin improved on older versions, first producing his in 1742 and continuing to tinker with it for two decades. His plans were public knowledge. He arranged for Robert Grace, a member of the Junto, to manufacture and market it as "the Pennsylvania Fireplace," though everyone associated the pot belly of the stove with its increasingly portly inventor.[31]

Franklin was motivated not so much by efficiency as by public health. As early as the founding years of the Junto, he assigned its members the task of "how may smoky chimneys best be cured." He assumed that ill health was in part caused by the cold drafts that fireplace chimneys sucked into cracks in house doors and windows. A system for cooking and heating that did not draw in cold air would make the house a healthier place. The heat was separated from the smoke by a series of iron plates, the smoke disappearing up the flue while the heat radiated out into the living spaces.[32]

What made Franklin's science modern was not its terminology, though we still use it, nor the lasting fame it brought him (for he is still the man who brought lightning down from the heavens), but the kind of tie one sees between the electrical experiments and the Pennsylvania Fireplace. Franklin's science was applied science—to borrow a much-abused advertising tag line, "better things for better living." His interest in science was inseparable from his interest in improving the standard of living. It was no puzzle that the scientific experimenter was the same man who provided for public hospitals, free public education, a free library, or sanitary streets and water. They are of a piece and can still be seen in the Clean Water and Clean Air acts on which American environmentalism rests today.

Franklin's fascination with combinations of science and technology was only one object of his "leisure." Another was "every part of our civil government . . . imposing some duty on me." He was given a "commission of the peace" by the lieutenant governor, named to the city's common council, and elected to the colonial assembly, all within three years of his retirement in 1748. He joined a delegation to renegotiate a treaty with the Delaware Indians. Despite the favor shown him by the proprietary government, he chaffed at the colonial administration's lack of preparations for war—a war that would begin in the backwoods of his own colony. It was then that he wrote *Plain Truth* and lobbied for a militia.[33]

Franklin did not realize that his "duty" to the colony would soon be tested by an even more profound political and military contest than King George's War. It was during the French and Indian War (1754–1763) that his patriotic sensibilities awoke. Historians disagree whether this was an American patriotism, and imperial pride, or the English in him, roused by the dire straits of the war, but once again it spurred his pen.

The war was not fought over religious differences, like the sixteenth- and seventeenth-century wars, or for honor and glory, like the wars of earlier years. It was fought for national interest, a rational calculation of the worth of empire. That is why it began at the place where Indian traders met Indians to exchange furs for guns—the forks of the Ohio River. The French sought to contain the bursting population of the Anglo-American empire's western settlements by girdling the frontier with forts. French regular troops, French Canadian militia, and their Indian allies would hold this line. But the English settlers would not be contained, and almost by accident the war erupted in the woods. "The French and Indian War was a far-flung affair, even in its origins." Both France and England determined, for the first time, that their colonies in America must take priority over their contest for dominion in

Europe. To the New World they sent regiments of their regular troops and their best officers.[34]

The opening rounds of the war occasioned Franklin's proposal for a union of the colonies. When commissioners of five of the colonies met at Albany in the summer of 1754, ostensibly to arrange for Indian alliances with the Iroquois of New York, Franklin was ready with a far-reaching plan to redo the structure of empire. Much of it anticipated the problems of self-government the American revolutionaries faced in 1776 and the solutions that American constitution drafters would fashion. In addition, his "Short Hints towards a Scheme for a General Union of the British Colonies on the Continent" presented to the conference on June 28, 1754, called for the dominion system that Britain actually developed for its colonies in the nineteenth century and remains the frame of the British Empire: "In Such a Scheme the Just Prerogative of the Crown must be preserved or it will not be Approved and Confirmed in England." But Franklin saw obedience as part of a bargain. The crown was to secure "the Just liberties of the People . . . or the Several Colonies will Disapprove of it and Oppose it." For him, the proper relationship in empire was not superior home country and inferior colony, but the flow of liberty, like electricity, all through the empire. Government existed to protect the rights of the governed.

In his 1754 draft, Franklin also addressed the same question that would become the most vexing threat to the new United States in 1776: union. "The Power of all the Colonies should be Ready to Defend any one of them with the Greatest Possible Dispatch." He had a plan for this as well. "Suppose then that One General Government be formed Including all the British Dominions on the Continent Within and Under which Government the Several Colonies may Each Enjoy its own Constitution Laws Liberties and Privileges as so many Separate Corporations in one Common Wealth."[35]

The particulars of the Albany Plan called for a grand council (much like the Senate under the federal Constitution—recall that state legislators chose U.S. senators until the Seventeenth Amendment provided for direct elections) "to Consist of two members at least Chosen by the Representatives of each Colony in Assembly." Elections were to take place every three years. (A "triennial bill" to provide for Parliamentary elections every three years was a perennial reform project in England, but never successful.) Over this council "a President General [would] be appointed by the Crown to Receive his Salary from Home." The royal salary would ensure that this new official was not beholden to the council for funds (royal governors' salaries were paid by their colonies' legislatures, which gave the legislatures great leverage over the royal governor). The president general would have the powers of royal governors (and something of the same powers that the Constitutional Convention of 1787 gave to the president of the United States).

The new union would be supported by imposts on each colony, according to its ability to pay, and "Therefore Let the Money arise from somewhat that may be nearly proportionable to Each Colony and Grow with it, Such as from Excise upon Liquors Retailed or Stamps on all Legal Writings Writs &c. or both to be Collected in Each Province and Paid to a Treasurer to be Appointed in each Colony by the Grand Council to be Ready on Orders from the President General and Grand Council." The proposed tax was very similar to a plan that George Grenville would introduce in Parliament in 1765, called then a stamp tax. It was rejected by the colonies. But the confederation of states during and immediately after the American Revolution faced the same problem, and the difficulty the new nation's government experienced in obtaining requisitions from the individual states would lead to the federal Constitution of 1787.

With war on the western frontier looming, the colonies had to find some way to coordinate their Native American policies with those of the home country. The council and president would "hold or order all Indian Treaties, Regulate all Indian Trade, make Peace and Declare war with the Indian Nations, Make all Indian Purchases of Lands not Within the Bounds of Particular Colonies, Make new Settlements on Such Purchases by Granting Lands Reserving a Rent for the General Treasury Raise and Pay Soldiers and build forts to Defend the frontiers of Any of the Colonies, Equip Guardships to Scour and Protect the Coasts from Privateers and Pirates, Appoint all Military officers that are to Act Under the General Command, the President General to Nominate and the Council to approve." These were among the "enumerated" powers the federal Constitution granted exclusively to the federal government. Franklin was again ahead of his time.

The plan was approved by the conference and sent to the various colonial legislatures for their assent. Not one of them agreed to it, even Franklin's Pennsylvania. He could have guessed as much—the assembly refused to discuss the matter of colonial union. The tax proposal was universally rejected. In addition, the Pennsylvania legislators were afraid, as were those of the

Benjamin Franklin, by Mason Chamberlin, 1762. Franklin the scientist, surrounded, somewhat imaginatively, by his apparatus. Philadelphia Museum of Art.

other colonies, that the plan gave too much power to the Crown or to other colonies—a jealousy that endangered federal union after the Revolution.[36]

Whitefield and Franklin crossed paths often enough during and after the French and Indian War—Whitefield visiting Franklin in Philadelphia and staying as a guest in Franklin's home, or Franklin meeting Whitefield in London. The last occasion the two came together was a momentous one. Franklin was the agent for the colony of Pennsylvania, an unofficial post created by the colony's assembly to represent its interests in Parliament (and a studied offense to the proprietor, Thomas Penn, who cordially hated Franklin). Although the post was entirely outside of the framework of official imperial governance, Franklin was sufficiently respected to be summoned to Parliament to explain why the colonies had erupted in violent protest against the Stamp Act of 1765. Whitefield, already an American in his loyalties, watched with approval as Franklin became America's spokesman.

The Enlightenment did not see the end of old forms of government, particularly monarchy. That belonged to the Age of Revolution beginning in America and spreading to the Continent. But Enlightenment reforms saw old forms of government and the economy in a new light. The key was the concept of enlightened self-interest. Governments could not exist without revenues. The raising of necessary funds was, however, never popular among the subjects of the Crown. To finance its own operation and to protect its empire at sea and on land, the British government had imposed a heavy tax load on Britons. There were duties on imports, taxes on hearths, excise taxes on goods for sale, and, most intrusively, stamp taxes. In the eighteenth century, one defense of these taxes was that they were in the self-interest of the taxed. Government provided essential services that benefitted everyone, and the general interest should supersede individual self-interest. The fullest exposition of this idea would come in the Scottish economist and philosopher Adam Smith's *Wealth of Nations* (1776). Smith argued that the individual should surrender his self-interest to the general interest or the public interest when it promoted long-term benefits for all. The "general interest of the country" would promote the particular interests of buyers and sellers, producers and consumers. He knew that the colonists would never trust this formula (after all, he published in 1776), but he supposed that if "the people on the other side of the water are afraid, lest their distance from the seat of government might expose them to many oppressions . . . their representatives in Parliament, of which the number from the first ought to be considerable, would easily be able to protect them from all oppression." In other words, the guardian of enlightened self-interest when it came to taxation was representation. Franklin favored colonial representation in Parliament and colonial self-taxation.[37]

The Stamp Act, passed in Parliament in April of 1765, was designed to raise revenue in America. Americans were required to use pre-embossed paper, supplied by ship from England, for a wide variety of official documents, including newspapers. The colonial agents had warned that such a direct tax on the colonists would be resisted. In February 1765, Franklin himself had tried to convince George Grenville, chancellor of the exchequer and the king's prime minister, not to present the act to Parliament—ironic in light of Franklin's inclusion of a very similar scheme of taxation in his Albany Plan of Union. But neither Grenville, who championed the act, nor anyone in Parliament (including the handful of members of the House of Commons who voted against the act), nor the colonists who agreed to serve as distributors of the stamped paper anticipated the virulence of the colonial response.

Colonial assemblies passed resolves against the legislation and warned that the delivery of the stamps in the fall would lead to violence. The warning was well founded. In city after city, violent protests erupted. The Grenville ministry and its appointees in America were appalled and affrighted at the vehemence of the opposition. Boston led the way. Samuel Adams, Paul Revere, and the other members of the "Loyal Nine," a group of merchants and craftsmen who organized the protests, set the mob in motion, and it pressured the new stamp collector Andrew Oliver to resign his commission by demolishing his place of business and besieging his house. When Lieutenant Governor Thomas Hutchinson tried to quell the demonstrations, he too became its target. On the evening of August 26, 1765, while at supper with his children, Hutchinson learned that the rioters were on their way to his house. Pulled away from his house by his daughter's plea that he depart, he was not there to see the mob break down his door with axes. The crowd surged through the house and the cellar and rushed back out into the street in search of Hutchinson. He returned at four in the morning to find that "one of the best finished houses in the province had nothing remaining but the bare walls and floors." They also carried off his tableware, family pictures, furniture, clothing, and about nine hundred pounds sterling in money and destroyed his books and papers, including "manuscripts . . . I had been collecting for 30 years besides a great number of public papers in my custody."[38]

On February 13, 1766, with the stamps on their way back to England or rotting in ships' holds in colonial harbors, Franklin stood in the well of the House of Commons and for four hours withstood a barrage of questions about the reception of the Stamp Act in the colonies—particularly Massachusetts. Some of the questions were genuine attempts to understand what had gone wrong. Franklin was not regarded as sympathetic to the riots, and rightly so, for he had proposed his friend John Hughes for the post of stamp collector, and Franklin's house, along with Hughes's, had been a target of the rioters in Philadelphia. His wife Deborah, her brother, and his friends mobilized to defend the newly finished building. Franklin heard about the tempest weeks later and wrote his wife, "I honor much the spirit and courage you showed."[39]

Grenville was one of the interrogators, but Grenville's bluntness having exhausted King George's patience, Charles Watson-Wentworth, the marquis of Rockingham, was now the leader of the government, and he was well disposed toward Franklin. Most of the questions were "scripted" by the new government, and Franklin probably knew in advance what they would be—after all, he had worked assiduously behind the scenes to convince Parliament to repeal the Stamp Act once the protests were known. Whitefield, watching from perhaps as close as the visitor's gallery, appreciated Franklin's performance, and as one actor to another he showered accolades on his collaborator: "Dr. Franklin has gained immortal honor by his behaviour at the bar of the House. His answer was always found equal to the questioner. He stood unappalled, gave pleasure to his friends, and did honour to his country."[40]

Parliament's (relatively) newfound reliance on expert testimony gave the event a modern feel, even without the C-SPAN and Twitter coverage one expects today. Franklin was asked to attend as an expert on American public opinion. Choosing Franklin as the expert made sense. Not only was he respected for his writings and his scientific achievements, but he was the postmaster general of the colonies. Grenville was a veteran bureaucrat with experience in treasury affairs, but he had never visited the colonies, nor had Rockingham. Franklin had his fingers on and in all manner of colonial matters. From his experience at the Albany conference, he knew the leaders of other colonies and had thought about intercolonial cooperation. He read other colonial newspapers. He was a veteran newspaper man, already busy trying to mold public opinion in Britain by writing letters to newspapers—"I hope . . . to see prudent measures taken by our rulers such as may heal and not widen our breaches."[41]

What was remarkable in his appearance before the Commons was not their choice of him, then, but their apparent reliance on his expertise, or any expertise for that matter. The increasing reliance on machinery in the economy of Britain, the rise of science as a field of knowledge, and the emergence of civil and mechanical engineering as professions all raised the visibility and importance of expertise. Experts were an increasingly common presence in English courts of law, testifying on their own knowledge about public works and private ventures.[42]

The concern for public opinion that one can see in the questions was just as important as Franklin's expertise. Public opinion was certainly not the basis by which the members of the House of Commons were chosen. That body, in theory the voice of the people, was in fact hardly representative of the English people. Only propertied adult male subjects of the king could vote, and before one voted, one had to profess loyalty to the Crown and acceptance of Protestant teachings. That left out over 90 percent of the adult male population. The rest were represented "virtually," the members of the lower house acting as trustees for the good of the whole—at least in theory. In fact, legislative politics was rife with corruption, the king buying up votes and various leading politicians routinely using their offices for profit and their patronage to repay political favors. The most appalling form of corruption was the maldistribution of seats in the house, with one uninhabited hill sending two members to Commons and Middlesex County, with a population of nearly one million, sending only eight men to Commons.[43]

But public opinion was finding its way into the great hall at Westminster where the Commons sat. Newspapers covered the debates. Members of Parliament gave newspapers copies of their speeches. Members also read the newspapers. Some members tried to "spin" newspaper accounts, including getting editors to change the report of Parliamentary activities. Perhaps even more indicative of the changing climate of politics, some editors were able to tell members of Parliament what they must say and do. To be sure, those out of Parliament had to be careful in how they tried to orchestrate public opinion, but such arrangements were more and more prevalent as the eighteenth century wore on.[44]

The questions asked Franklin hinted that the House of Commons recognized that public opinion in the colonies was essential to the success of any legislative program. In most of the colonies by 1765, just about all white males who paid taxes, owned property, or served in the militia could vote. Proportionally, ten times as many colonists could vote for members of their legislators as Englishmen could vote for members of the Commons. In many of the colonies, voter deference to local elites determined who won contests for the assembly, but as the Stamp Act riots demonstrated, elite status did not matter when the public was aroused. Ordinary men could lead the mob, and men of property and standing could be its victims. In addition, the flood of petitions from the colonies against the Stamp Act and the resolves of the extralegal Stamp Act Congress that met in New York City in October 1765 showed that colonial public opinion could be marshaled against acts of Parliament with surprising ease and effectiveness. Parliament had received and rejected these petitions in 1765, and perhaps as the members gathered to hear Franklin, they regretted their haste.[45]

Franklin arranged for shorthand notes by observers and a verbatim account of the grilling to be published in Philadelphia—he knew as well as or better than his interrogators the importance of public opinion. But the resulting account of the hearing had a twist characteristic of all of Franklin's writings. By laying the very terse questions of the members of Commons alongside his longer, fuller answers, he gave the reader the (intended) impression that he was lecturing the legislators rather than being interrogated. The result: from the very first question, he starred. "Q[uestion]. What is your name, and place of abode? A[nswer]. Franklin, of Philadelphia." An astute American reader would take the correct inference from the first Q and A that Franklin as a scientist had made his name a household word in the home country.[46]

The members of Commons assumed that Franklin could speak for the colonists. In a way, this showed a profound ignorance of the wide variety of peoples and associations in the colonies. For his own part, Franklin did not soften the impact of his answers with such nuance.

Q. Do the Americans pay any considerable taxes among themselves?

A. Certainly many, and very heavy taxes.

Q. What are the present taxes in Pennsylvania, laid by the laws of the colony?

A. There are taxes on all estates real and personal, a poll tax, a tax on all offices, professions, trades and businesses, according to their profits; an excise on all wine, rum, and other spirits; and a duty of Ten Pounds per head on all Negroes imported, with some other duties.

Q. For what purposes are those taxes laid?

A. For the support of the civil and military establishments of the country, and to discharge the heavy debt contracted in the last war.

In short, the colonists were paying for their own debts, taxing themselves. They were not shirkers, tax evaders, or insensible of the needs of government.

The members of Commons persisted, driven by the new Rockingham ministry's need to raise a revenue in the colonies without spurring widespread resistance. "Q. How long are those taxes [the colony imposed on itself] to continue? A. Those for discharging the debt are to continue till 1772, and longer, if the debt should not be then all discharged. The others must always continue. Q. Was it not expected that the debt would have been sooner discharged? A. It was, when the peace was made with France and Spain—But a fresh war breaking out with the Indians, a fresh load of debt was incurred, and the taxes, of course, continued longer by a new law." Franklin referred to what is now called Pontiac's rebellion, a league of western Indians formerly allied to the French who, when the French surrendered Canada and New France (their colony between the Appalachians and the Mississippi River) to Britain in 1763, refused to accept peace.

As time passed, the questions grew more pointed, the riots casting their shadow on the interrogation. "Q. Are not all the people very able to pay those taxes? A. No. The frontier counties, all along the continent, having been frequently ravaged by the enemy, and greatly impoverished, are able to pay very little tax. And therefore, in consideration of their distresses, our late tax laws do expressly favour those counties, excusing the sufferers; and I suppose the same is done in other governments."

Franklin was offering some special pleading, which some members of the lower house might have guessed. For hidden in the ongoing controversy over Indian policy was a competition among land speculation companies to exploit the trans-Appalachian region opened to speculation by the 1763 Treaty of Paris. Rival companies organized in Britain and the colonies vied for this plum, and Franklin was associated with a number of the American schemes. Franklin had even revealed something of his machinations in this regard to Whitefield ten years earlier:

I sometimes wish, that you and I were jointly employ'd by the Crown to settle a Colony on the Ohio. I imagine we could do it effectually, and without putting the Nation to much Expence. But I fear we shall never be call'd upon for such a Service. What a glorious Thing it would be, to settle in that fine Country a large Strong Body of Religious and Industrious People! What a Security to the other Colonies; and Advantage to Britain, by Increasing her People, Territory, Strength and Commerce. Might it not greatly facilitate the Introduction of pure Religion among the Heathen, if we could, by such a Colony, show them a better Sample of Christians than they commonly see in our Indian Traders, the most vicious and abandoned Wretches of our Nation?[47]

The questioning then became more personal. "Q. Are not you concerned in the management of the Post-Office in America? A. Yes. I am Deputy Post-Master General of North-America. Q. Don't you think the distribution of stamps, by post, to all the inhabitants, very practicable, if there was no opposition? A. The posts only go along the sea coasts; they do not, except in a few instances, go back into the country; and if they did, sending for stamps by post would occasion an expence of postage, amounting, in many cases, to much more than that of the stamps themselves." Franklin had for the moment dodged the question of whether he would in any way hinder the distribution of the stamps.

Grenville could no longer remain silent. With his entry into the proceedings, they began to take on the shape of a boxing match. "Q. Are not the Colonies, from their circumstances, very able to pay the stamp duty? A. In my opinion, there is not gold and silver enough in the Colonies to pay the stamp duty for one year. Q. Don't you know that the money arising from the stamps was all to be laid out in America [Grenville's justification for the tax when he proposed it]? A. I know it is appropriated by the act to the American service; but it will be spent in the conquered Colonies, where the soldiers are, not in the Colonies that pay it." The round went to Franklin on points.

The probing continued, this time focusing on Pennsylvania. "Q. What number of Germans [were there in the colony]? A. Perhaps another third; but I cannot speak with certainty. Q. Have any number of the Germans seen service, as soldiers, in Europe? A. Yes,—many of them, both in Europe and America. Q. Are they as much dissatisfied with the stamp duty as the English? A. Yes, and more; and with reason, as their stamps are, in many cases, to be double." Round two went to Franklin, with a technical knockout, for he had special knowledge of the case.

Economic questions followed. Pennsylvania had an unfavorable balance of trade with the home country, that is, the value of its imports exceeded that of its exports. It thus owed money to British merchants.

Q. How then do you pay the balance?
A. The Balance is paid by our produce carried to the West-Indies, and sold
   in our own islands, or to the French, Spaniards, Danes and Dutch; by the

same carried to other colonies in North-America, as to New-England, Nova-Scotia, Newfoundland, Carolina and Georgia; by the same carried to different parts of Europe, as Spain, Portugal and Italy. In all which places we receive either money, bills of exchange, or commodities that suit for remittance to Britain; which, together with all the profits on the industry of our merchants and mariners, arising in those circuitous voyages, and the freights made by their ships, center finally in Britain, to discharge the balance, and pay for British manufactures continually used in the province, or sold to foreigners by our traders.

Franklin had succinctly explained how the Navigation Acts or "mercantilism" worked, favoring the home country and forcing the colonial merchants to scurry about to make ends meet. Pennsylvanians were helping make Britain rich. But the round was scored even.

To concede that the purpose of the Navigation Acts was to keep the colonies economically dependent on the home country, as Franklin had stressed in reply to his questioners, was hardly an appetizing morsel for them to swallow. Instead, they turned to national security.

Q. Do you think it right that America should be protected by this country, and pay no part of the expence?

A. That is not the case. The Colonies raised, cloathed and paid, during the last war, near 25000 men, and spent many millions.

Q. Were you not reimbursed by Parliament?

A. We were only reimbursed what, in your opinion, we had advanced beyond our proportion, or beyond what might reasonably be expected from us; and it was a very small part of what we spent. Pennsylvania, in particular, disbursed about 500,000 Pounds, and the reimbursements, in the whole, did not exceed 60,000 Pounds.

Round four went to Franklin on punch count.

And here was the climactic question: "Q. Do you not think the people of America would submit to pay the stamp duty, if it was moderated? A. No, never, unless compelled by force of arms." There were those in the colonies who thought Franklin too friendly to Parliamentary impositions like the Stamp Tax. This was his answer to them and to Parliament. A direct tax, levied without colonial representation in Parliament, would never be accepted in America. Franklin was not a lawyer and did not think in legalistic terms.

He did not say that such a tax would violate the English Constitution, as some American protestors (notably lawyers such as John Adams, Daniel Dulany, James Wilson, John Dickinson, Patrick Henry, and Thomas Jefferson) did. Franklin simply said that no such tax was feasible. He had won the bout.

As a matter of law, however, the question of constitutional obligations is still debatable. There was no fundamental English constitution in the sense of the U.S. state constitutions or the federal Constitution. After the ouster of King James II in 1688, Parliament claimed and William and Mary conceded that Parliament would be the supreme authority in England. Thus, if Parliament imposed a tax on English colonies, it was prima facie constitutional. But if Parliament itself was controlled by its own ancient usages and its declarations of rights, then it could not do something that violated those usages. So argued Massachusetts lawyer James Otis Jr. in the famous writs of assistance cases of 1761. That became the basis of the colonial protest against "taxation without representation" in 1765 and 1766.

In addressing these kinds of questions, the interrogation of Franklin shifted from a boxing match to a genuine conversation about the legal nature of the imperial relationship. It foreshadowed the questions that the ministry of Lord North and the opponents of parliamentary governance of the colonies would raise in coming days. Did disobedience to the Stamp Act demonstrate a deeper disloyalty to Britain? Was it the opening wedge of treason?

Q. What was the temper of America towards Great-Britain before the year 1763?

A. The best in the world. They submitted willingly to the government of the Crown, and paid, in all their courts, obedience to acts of Parliament. Numerous as the people are in the several old provinces, they cost you nothing in forts, citadels, garrisons or armies, to keep them in subjection. They were governed by this country at the expence only of a little pen, ink and paper. They were led by a thread. They had not only a respect, but an affection, for Great-Britain, for its laws, its customs and manners, and even a fondness for its fashions, that greatly increased the commerce. Natives of Britain were always treated with particular regard; to be an Old England-man was, of itself, a character of some respect, and gave a kind of rank among us.

Q. And what is their temper now?

A. O, very much altered.

In short, Parliament had profligately rifled that store of goodwill.

At this point, well into the questioning, members of Parliament were thinking of remedies. Could the Stamp Tax be repealed without too much loss of face? In fact, a little less than three weeks later, the Rockingham ministry would seek and gain Parliamentary approval for the repeal. "Q. Did you ever hear the authority of Parliament to make laws for America questioned till lately? A. The authority of Parliament was allowed to be valid in all laws, except such as should lay internal taxes. It was never disputed in laying duties to regulate commerce." Here Franklin may have laid the way to the next round of Parliamentary revenue-raising schemes—new customs duties. Whether he made this concession (for it was not necessary to answer the question) to gain the goodwill of the ministry and open a door to his own preferment or simply because he understood that an impoverished England was not good for the colonies cannot be determined. The fact is that he did seek preferment and would have accepted it.

> Q. In what light did the people of America used to consider the Parliament
>   of Great-Britain?
> A. They considered the Parliament as the great bulwark and security of their
>   liberties and privileges, and always spoke of it with the utmost respect
>   and veneration. Arbitrary ministers, they thought, might possibly, at
>   times, attempt to oppress them; but they relied on it, that the Parliament,
>   on application, would always give redress. . . .
> Q. And have they not still the same respect for Parliament?
> A. No; it is greatly lessened.
> Q. To what causes is that owing?
> A. To a concurrence of causes; the restraints lately laid on their trade,
>   by which the bringing of foreign gold and silver into the Colonies was
>   prevented; the prohibition of making paper money among themselves;
>   and then demanding a new and heavy tax by stamps; taking away, at the
>   same time, trials by juries, and refusing to receive and hear their humble
>   petitions.

The situation was a grave one, but not without hope.

Franklin, always practical, always looking for solutions to problems, ready to manage rather than to fight, offered a compromise: "Q. Then no regulation with a tax would be submitted to? A. Their opinion is, that when aids to the Crown are wanted, they are to be asked of the several assemblies, according to the old established usage, who will, as they always have done, grant them freely. And that their money ought not to be given away without their consent, by persons at a distance, unacquainted with their circumstances and abilities." For the empire to survive, the metropolitan authorities must recognize the growing autonomy of the colonies and base taxation on the consent of the colonists. "They think it extremely hard and unjust, that a body of men, in which they have no representatives, should make a merit to itself of giving and granting what is not its own, but theirs, and deprive them of a right they esteem of the utmost value and importance, as it is the security of all their other rights."

At last the crucial issue came out of the shadows. It was a question of privileges versus rights. The language of rights was new, and it was important that the Parliament understand it. The first colonists had believed that liberty lay in the privileges that royal grants, charters, and other documents conferred. Thus, the liberty to create and maintain local legislatures was a privilege derived from colonial charters. A settler could claim certain so-called liberties—the right to follow a trade or vote in an election—as a privilege granted by law or custom. Good law enumerated such liberties so that everyone knew where he or she stood. Property for the first colonists had been inseparable from privilege, because the common law allowed a person to enjoy one's land and chattels. But every privilege could be rescinded by the same power that granted it. Even ownership of land in the colonies could be revoked, for the privilege of private property in the king's American domains was not absolute. Thus, petitions to the Crown invariably referred to the colonists' "privileges."[48]

In quite contrary fashion, the American protestors had come to think of liberty as a natural quality of life, a right, that government could not constrain or deny. For the protestors, guarantees of fair trial before local juries, barriers to illegal searches and seizes, and private property ownership were all rights that government could not diminish, and that included taking property by taxation without the consent of the owners. In the continuing constitutional convention that would be the revolutionary era, American political theorists even began to assume that one of the foundational purposes of government was to protect rights rather than to grant and define privileges.[49]

The questioning went on for hours, and the members of Parliament had reached as far into Franklin's knowledge as they could. The exchange ended with a whimper.

Q. If the stamp-act should be repealed, and an act should pass, ordering the assemblies of the Colonies to indemnify the sufferers by the riots, would they obey it?

A. That is a question I cannot answer. . . .

Q. Then may they not, by the same interpretation, object to the Parliament's right of external taxation?

A. They never have hitherto. Many arguments have been lately used here to shew them that there is no difference, and that if you have no right to tax them internally, you have none to tax them externally, or make any other law to bind them. At present they do not reason so, but in time they may possibly be convinced by these arguments. . . .

Q. What used to be the pride of the Americans?

A. To indulge in the fashions and manufactures of Great-Britain.

Q. What is now their pride?

A. To wear their old cloaths over again, till they can make new ones.

Franklin foresaw, correctly, that the protestors would become advocates of homespun. Even George Washington, whose clothes chest included brocade, would don "old cloaths" when necessary. In 1765 he wrote to a London merchant that Americans might have to do without some of the finery of dress if they were to be seen from England as an independent-minded people.[50]

The glory of empire in American eyes had not departed, but it had dimmed. Would a final crisis bring the empire or the colonists to their knees? No one could tell. Franklin returned to his Craven Street abode tired but satisfied and hopeful. He was still a healthy sixty years old and still thought that his future lay with Britain's empire. He would remain in England until 1773, praying for a reconciliation between Britain and its colonies, but increasingly convinced that there would be no middle ground, no basis for compromise. Sometime in late summer 1769, he wrote to Whitefield, "I am under continued apprehensions that we may have bad news from America. The sending soldiers to Boston always appeared to me a dangerous step; they could do no good, they might occasion mischief. When I consider the warm resentment of a people who think themselves injured and oppressed, and the common insolence of the soldiery, who are taught to consider that people as in rebellion, I cannot but fear the consequences of bringing them together. It seems like setting up a smith's forge in a magazine of gunpowder."[51]

As for religion, Franklin had the last word in a loving letter to Whitefield. It combined the skepticism of the *Dissertation* with a far wiser sense of imperial affairs. "I rather suspect, from certain circumstances, that though the general government of the universe is well administered, our particular little affairs are perhaps below notice, and left to take the chance of human prudence or imprudence, as either may happen to be uppermost. It is, however, an uncomfortable thought, and I leave it."[52]

Whitefield kept Franklin in his thoughts and prayers. On January 21, 1768, he wrote to Franklin,

> Through rich grace I can sing "O Death where is thy sting"—but only through Jesus of Nazareth. Your Daughter I find is beginning the world. I wish you joy from the bottom of my heart. You and I shall soon goe out of it—Ere long we shall see it burnt—Angels shall summon us to attend on the funeral of Time—And (Oh transporting thought!) we shall see Eternity rising out of its ashes. That you and I may be in the happy number of those who in the midst of the tremendous final blaze shall cry Amen—Hallelujah.

He had not given up hope that Franklin would see the light that grace alone could shine on a soul. Whitefield returned to America in 1769, much older than his scant fifty-five years. His wife had passed away; his always uncertain health was failing him. He had become something of a "Whig," pleading for the American cause. Privately he lamented the demise of American liberties and the selfish destructive policies of a Parliament gone wild. He threw himself into another tour of preaching, exhausting his slender reserves, and in Newburyport, they gave out. He had come home, to America. And there he died.[53]

Epilogue
The Birth of the Modern World

FOR MOST AMERICANS TODAY, modern means scientifically advanced and technologically sophisticated, along with improved standards of living, mastery of nature, and a society that rewards individual initiative and creativity. It is a kind of faith, according to the philosopher Stephen Toulmin, that we can make the world better by applying ourselves to the task. This was the future that Benjamin Franklin envisioned for America two and a half centuries ago.

But alongside that brand of secular, material thinking exists another worldview whose millions of adherents long for a sacred moral purity. Theirs was and still is a spiritual fellowship whose passion for salvation one can see in the faces of the worshipers at any evangelical church on a Sunday morning. Born-again Christians are a force in American cultural and political life. They are the inheritors of the revivalism that George Whitefield brought to America in 1739.

While cultural observers routinely note how different these two cultures are today, indeed how they seem to be polar opposites, in fact they share much more than a surface comparison shows. Both arose in the provincial setting of British North America. Both depended on the transformative effects of newly developed mass media and the assumed equality of the consumers of such media. Both drew energy and direction from the efforts of charismatic leaders.

No one would claim that Franklin and Whitefield were prototype modern men, much less that they originated the Web or espoused democracy in its modern sense, but both Franklin's technology and Whitefield's revivalism thrived on mass production of the printed word, the driving force behind modern culture. From Franklin's "Silence Dogood" newspaper pieces of the 1710s and Whitefield's *Journals* of the 1740s to the blogs, e-books, and egalitarian culture of today seems a mighty leap in time and space, but the foundations of modern media, of promotional literature and advertising, and of the belief that all people, regardless of class, caste, or origin, were entitled to consume the products of that media were first envisioned and popularized by these two very special men.

Franklin and Whitefield were not modern men. They had no sense that they were creating the foundations for our world. But in remarkable and important ways, their values and contributions paralleled, contributed to, and in some cases were largely responsible for ideas and values we exhibit; anticipators, forerunners, prototypes, they set the wheels of modernity in motion.

The similarities between then and now that Franklin and Whitefield exhibited and encouraged but that did not exist in any distinct form previously are numerous. Franklin seems the more modern in terms of parallels. Franklin's forward-looking versatility is strikingly modern. Blogger, bureaucrat, grant writer, policy "wonk," electronic tinkerer, and media mogul all fit together in his remarkably innovative and inquisitive lifestyle. One can almost see him at work at a computer on his blog, much as one might see him in the 1730s, setting his own little essays in print for the *Gazette*. No modern bureaucrat worked harder in his office than Franklin did in his, preparing grant applications or laying out policy recommendations. Franklin and his "associates" in their makeshift lab were the forerunners of the "basement and garage inventors" today. Most striking, perhaps, is how much Franklin's media empire resembles that of the modern moguls, the Rupert Murdochs and their ilk.

Franklin was not a democrat in the modern sense of the word. He had bondmen and bondwomen servants and at one time some dealings with the slave trade. He hated Catholics, looked down on Indians, and did his share of racial sneering. In the 1770s, he condemned slavery—chattel slavery in the colonies, colonial enslavement to a tyrannical Parliament—but he did not go public with a wholesale antislavery program. He confined his remarks to private correspondence with antislavery men. But he did espouse a kind of equality—the equality of thinking classes, an equality of liberal-minded men that knew no caste or class. He opposed the system of privilege by birth, or privilege conferred by something other than a person's merit and abilities. He chaired the committee that prepared the first Pennsylvania state constitution, with its ringing (if somewhat derivative) language that all men were "born equally free and independent," and unlike the author of the Declaration of Independence, he proposed that all men should have the right to vote and hold office. In fact, Franklin served with Thomas Jefferson on the committee that wrote the Declaration of Independence, and he may have contributed something to the formulation of its ringing introductory passages. In any case, Franklin's was the equality of the modern meritocracy, a system of equality of access to an education, the forerunner of the public college and university systems we enjoy today.[1]

Whitefield's views paralleled the contrapuntal force in modern American culture to secular progress, an evangelical longing for a supposedly purer past, much as Whitefield longed for a primitive Christianity. As historian Christopher Lasch wrote at the end of the twentieth century, "The assumption that our standard of living . . . will undergo steady improvement colors our view of the past. It gives rise to a nostalgic yearning for a bygone simplicity." That yearning finds comfort in evangelical religion. Today (and likely in the future) polls tell us that the United States is the most religious country of all the advanced, wealthy nations—and that by far. No movement so perfectly demonstrates this longing as the "search for a Christian America." Led by evangelicals, this reaction to secular humanism "becomes an appeal to recover the Christian roots, the Christian heritage, the Christian values of an older America." While such an effort may require that one don blinders to the Christian defenses of slavery, imperialism, and intolerance also characteristic of "older America," one cannot deny that the evangelical preaching of the eighteenth century is a centerpiece of this movement, with George Whitefield in the van.[2]

Comparisons between Whitefield and today's revivalists may seem forced, but when one focuses on their use of the media, the parallels are striking. Whitefield shared his tour stops, stories, sermons, and journals with newspapers and publishers in England and America. He never missed the opportunity to use every available media outlet to advance his ministry. Today evangelicals follow a similar strategy. Televangelism (a neologism coined by *Time* magazine) is a modern, American phenomenon, with channels on cable TV devoted to nonstop revival preaching, but Whitefield would have been a marvel on television, his dramatic presentation, his vocalizations, even his gestures magnified by the media and brought into homes with perfect clarity (though he would have required eye exercises or surgery to correct his lazy eye problem). Whitefield's itinerancy is also mirrored in modern revival tours, starting with Billy Sunday's and Aimee Semple McPherson's in the 1920s and continuing with Billy Graham's. Some televangelists face the same kind of criticism as Whitefield for deviating from accepted Christian doctrine. Whitefield's attempts to raise money for the Bethesda orphanage likewise resemble the efforts that the more scrupulous televangelists use to raise funds for their ministries. Whitefield never converted the funds to his personal use, however, unlike some of the televangelists.

The modern evangelical revival is open to all who would come; it makes no distinctions based on wealth or status. It is egalitarian in its program, just as Whitefield's preaching in the fields and town squares was open to all. Though he would find himself perfectly at home in the mansions of the noble and the rich, he still stopped to preach to the wayfarer and the downtrodden. In short, he helped create and his ministry in turn was buoyed by the "egalitarian forces powerfully at work within popular religious culture."[3]

Whitefield was the ultimate democrat in a time of rank and station. His audiences, though primarily middle class, were open to everyone. Men and women attended. Rich and poor stood silently next to one another. "When the reverend George Whitefield preached in Boston, in the Fall of 1740, people from all walks of life—rich and poor, black and white, free and unfree, young and old, men and women—flocked to hear his powerful message." For him, and in his presence, "Christ's hearers" knew no distinctions of rank and gender, or even of race (in New England at least). Whitefield's revival, continued after he departed, filled Boston churches with all comers.[4]

A final parallel: Franklin's and Whitefield's fascination with the printed word is thoroughly modern. Though both men still lived in societies in which communication was primarily face-to-face and oral, both spent an immense amount of time preparing words for the printed page. Franklin was ever writing for publication, though sometimes prudence dictated that he not publish. Whitefield undertook a thorough revision of his journals and sermons for a new print edition in 1757. Today the printed word arrives in e-books, on screen, and in handheld electronic devices, but it is still the printed word.

The contributions to and connections between Franklin's and Whitefield's activities and those of today's world are a little harder to trace than the parallels, so much time having passed between the eighteenth and the twenty-first centuries. Nevertheless, they developed in their thinking and their conduct an approach to life that looked forward to our own.

For Franklin, these connections included "Yankee" values, technological savvy, and the espousing of American exceptionalism. In the nineteenth century, Franklin was celebrated as the essential Yankee—hard working, thrifty, skeptical of unproven claims, and innovative. Defenders of traditionalism attacked Yankeedom as shallow and materialistic, but as Esmond Wright argued, "it was these Yankee values that were transforming the eighteenth-century world" into one that made nineteenth-century American capitalism possible. If this Yankee appellation was something of a caricature that Franklin would have found both amusing and unfair, Franklin's appearance in modern advertisements for banks and financial services demonstrates the continuing effectiveness of this identification.[5]

The Yankee is a booster, a civic promoter, extolling town life as he improves it. Franklin was the very model of a booster. The water works, the street lights, and the other civic improvements he urged for Philadelphia are testaments to his Yankee boosterism. As Carl Van Doren wrote in his prize-winning biography of Franklin, "No other town, burying a great man, buried more of itself than Philadelphia with Franklin."[6]

The Yankee was a joiner. No longer bound to a particular place or station in life, he was free to volunteer for membership in groups. Indeed, one scholarly distinction between modern and traditional societies is a person's capacity to decide which groups to join and which to shun. Franklin became a Mason, founded the Junto, opened the American Philosophical Society to public membership, and fostered an ideology of volunteerism. This ideology rested on two principles: individual choice and equality. Franklin chose which church to attend, or not to attend. He decided for himself whether to travel to England or remain home. Americans abroad, he believed, were equal to Englishmen at home. Members of the Junto were equal to one another. Names, family, wealth did not matter. Only the decision by the individual to join and participate mattered. Electricity affected everyone equally. So, too, in politics, when it came to the colonists' rights and the Crown's pretensions,

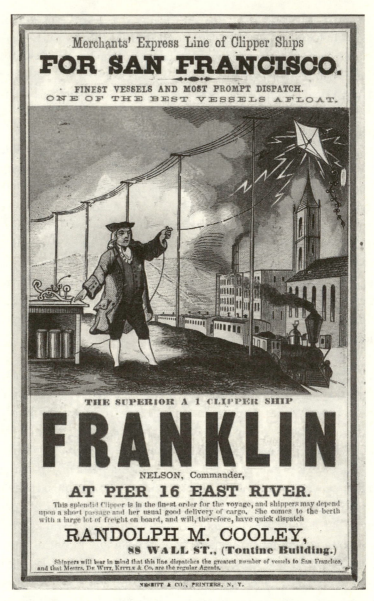

S.S. Benjamin Franklin. By the mid-nineteenth century, Franklin had become a symbol of American energy, enterprise, and ingenuity. American Antiquarian Society.

"there is one above who rules these matters with a more equal hand." Neither individualism nor equality was a dominant theme in Western life at the beginning of the eighteenth century. At its end, with Franklin often cited as an example, both individualism and equality were synonymous with America.[7]

In a more general fashion, Franklin has come to symbolize the American—bigger, taller, more robust, more optimistic, and more capable than the European. Franklin actually promoted this image when he served the Continental Congress as a diplomat in France. At one now-famous dinner party in 1782, he told his French hosts that American air, food, and life were healthier than Europe's, and that was why America had escaped, thus far, European degeneracy. Franklin had planted the seeds of a potent version of Americans' view of themselves as an exceptional people. American exceptionalism, as the theory is now termed, is much debated among modern scholars, but the fact that Franklin advocated it cannot be doubted.[8]

Franklin's commitment to betterment of material life is our own. Every experiment, every theory he proposed, was tied in some fashion to practical concerns. Heat, light, energy, nutrition, health, and longevity were some of the main subjects of the Junto, the Philosophical Society, and all the other institutions that he planned, supported, and was proud to help govern. We no longer repeat Franklin's dangerous experiments, but insofar as many of them revealed the secrets of electricity, we owe a genuine debt to Franklin.

The connections between Whitefield's emotional preaching and that of our own day are clear. His extemporaneous, open-air, born-again ministry set a precedent for the methods of preachers in the "Second Great Awakening" of the early nineteenth century. "Local revivals yielded a rich harvest of young men, who once touched by God were ready to join the ranks of circuit riding preachers." Like Whitefield, they carried the evangelical message with them as they traversed the countryside. For example, it was unexceptionable for Charles Grandison Finney, the foremost of the Second Great Awakening's preachers in the nineteenth century, to employ dramatic gestures because Whitefield had made them acceptable on the pulpit. "In the fields and public squares of the cities, Whitefield was a fitting symbol of the noncreedal revivalism that characterized much of the revivalism of the early republic. Whitefield's name was also emblematic of the dramatic, extemporaneous style of preaching—culminating in the call for repentance and faith in Christ—that marked Methodist, Baptist, and evangelical Presbyterian [and] Congregationalist evangelistic services." He had become the model for the populist preacher. Though parodied and pilloried in works such as Sinclair Lewis's *Elmer Gantry*, the camp meeting anticipated by Whitefield's dramatic theatrics remains a feature of American evangelical life.[9]

The democracy of the First Great Awakening would reappear in the democracy of the revolutionary crowd, lifted to patriotic emotion by political pleading remarkably similar to that of the itinerant revivalist. Whitefield joined in this, becoming an ally of Old Lights like Boston's Charles Chauncy as they preached against the supposed tyranny of Parliament. "By the time of the American Revolution, the warmth of such evangelical appeals, and their ability to draw the unchurched into cohesive fellowships," worked in both political and religious spheres.[10]

Finally, there are the strongest, causal ties that link the two men to modernity. Franklin was the father of modern science. He did not invent the scientific method (that plum belongs to the English jurist and philosopher Francis Bacon), but Franklin moved science from the closet to the family room. Instead of an arcane and specialized field for a handful of men whose patrons were kings and nobles, Franklin's science was for the masses. He was the forerunner of the modern research grant writer and the public scientist, such as the astronomer Carl Sagan and the naturalist David Attenborough one sees today on PBS's *Nova* and other mass media.

What was more, his science, like ours, was entirely secular. Even Newton believed in ghosts, supernatural spirits, and magic. Franklin would have none of it. Franklin proved that electricity, in particular lightning, was not some divine punishment. It was merely a big version of the static electricity he produced, and controlled, in his experiments. While many scientists are religious, as a rule they do not bring their religious convictions into their laboratories. They accept evolutionary biology, a natural theory of the origin of the universe, and quantum mechanics theory in particle physics.[11]

Indeed, it is remarkable how much of Franklin is still with us, not just the images, but the inventions and the spirit behind them. The Library Company of Philadelphia thrives, a magnet for serious students of early America. Father Abraham still circulates, and the compilation of Poor Richard Saunders's adages rings true in the twenty-first century, even though Franklin's tongue-in-cheek manner goes unrecognized by most of the readers of *The Way to Wealth*. He was directly responsible for the College of Pennsylvania and the Pennsylvania Hospital, and the college he helped found now occupies much of West Philadelphia. Now named the University of Pennsylvania,

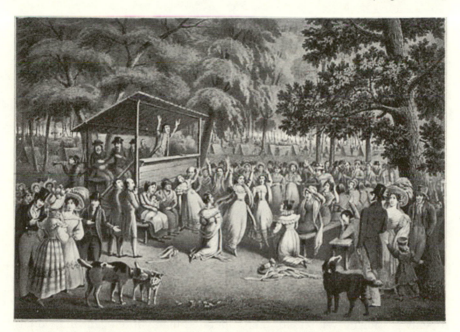

A typical camp meeting during the "Second Great Awakening" of the early to mid-nineteenth century. Its enthusiasm matched that of Whitefield's tours. By H. Bridport. Library of Congress.

it is a world leader in higher education. The author of this book spent many weeks at the Pennsylvania Hospital while his frequent coauthor and wife, Natalie Hull, recovered from back surgeries. The facade, the facilities, and the services are much improved since Franklin's day, but the heart of his contribution to modern life still beats on Spruce and Eighth.

Whitefield too is still with us. He is not so well known today as Franklin. Indeed, a poll of history students at the University of Georgia revealed that everyone knew Franklin and no one knew Whitefield. But one could argue that Whitefield was also a founder of higher education in America. Every one of the first colleges in America had a religious affiliation. Harvard was Congregationalist, Yale Presbyterian, the College of William and Mary Church of England. Whitefield raised funds for what would become Princeton, Dartmouth, and Penn. He had planned for a college in Georgia, and those plans came to fruition with the founding of the University of Georgia in 1785.

Sunday morning on the roads in Northeast Georgia reminds us that Whitefield's rhetorical contributions are still with us. Evangelists in megachurches, parking lots filled to bursting, broadcast his message of sin and repentance to hundreds of thousands of viewers in their living rooms. The word has become electronic, but it is still the Word. At its core, the doctrine of rebirth, the essence of evangelical religion, has kept Whitefield's message and its medium alive.

The partnership of equals between Franklin and Whitefield changed their world and ours. Two men on the rise found one another useful. Whitefield was already a phenomenon and Franklin was already on his way to wealth, but both men had a keen eye for advantage. Such partnerships are as common today as they were then. The partnership of equals promoted Whitefield in America and profited Franklin. The arrangement increased the prominence of the two, Franklin becoming an unofficial publicist for the revival, Whitefield boosting Franklin's visibility among the evangelicals.

Their partnership also reminds us that behind the public personas that both men cultivated, the images of the affable and frugal Franklin and the openly passionate Whitefield, there operated highly intelligent individuals. Both looked to the future rather than living in the moment. The prescriptions they laid down for the conduct of themselves and others were not a sham, and neither were they naive. Both Franklin and Whitefield knew the sin of pride and struggled to contain it. The moralism of the essays and the sermons was not just meant for the reader; it was self-constraining. The greater their fame, the harder the two men worked to prevent pride from escaping the paper walls, hence the almost compulsive need to publish.

In 1739, neither man would have dreamed of the severing of ties between the thirteen colonies and Great Britain. They were proud to be Britons and proud of the empire. They traveled back and forth between colonies and home country, even though their luck in the crossings was bad (they always seemed to be caught in storms or delayed by war). When both men, in the 1760s, protested against Parliamentary exactions, they spoke and wrote as Britons, advocating British liberties. To make them Americans is not to deny any of this, for American did not yet mean non-Briton.

They watched with mounting horror as the government of the home country, whose wealth was turning into luxury and whose leaders were falling to corruption, did not choose to follow the self-denying course that Franklin and Whitefield set. Whitefield would die before that sin of political arrogance earned its due reward. He passed on September 30, 1770, and was buried under the altar of the Old South Presbyterian Church of Newbury-port, Massachusetts, a hotbed of revivalism in the 1740s, and later a hotbed of patriot protest against the British Parliament. Whitefield never abandoned his allegiance to the Church of England, and it is surely ironic that he rests with the Puritan opponents of Anglicanism.

In these years Franklin sought reconciliation between the estranged colonies and the home government, and perhaps the governorship of Pennsylvania for his efforts. Denied this, he played a key role in stirring colonial anger against the home government and returned to America a revolutionary leader. With that Revolution, America became a confederation of independent republics, later a democratic republic and a beacon of reform to the nations of the world. Franklin died on April 17, 1790, full of honors, including service in the convention that drafted the federal Constitution. It is now the longest-existing frame of government in history. His remains rest in the Christ Church burying ground in Philadelphia, a final irony for a man who was reared in Puritanism, in later life became a deist, and never subscribed to the doctrines of the Church of England.

The faith that both Franklin and Whitefield had in their ideas is now under attack. The reassuring American concept of modernity based on an ever-growing material wealth and ever-renewing inspirational faith may have reached a point of diminishing returns. As Stephen Toulmin wrote, some years ago, "What looked in the nineteenth century like an irresistible river has disappeared in the sand . . . far from extrapolating confidently into the social and cultural future, we are now stranded and uncertain of our location. The very project of modernity thus seems to have lost momentum."[12]

Certainly the ideological certitudes of the Franklins and Whitefields are under assault in the recurrent waves of culture wars, ethnic wars, and imperial wars that tear at our social and intellectual fabric. A postmodern, apocalyptic vision is a staple of our movies and literature. The technologies that were to free us have instead kept us hostage to hackers and spammers, and religious belief once again has turned to harsh judgments of those who are not among the saved. Just as the closing years of the provincial era clouded Americans' view of the future, so ours has become unfocused. In this moment of uncertainty, perhaps our best course is to remind ourselves of the lessons of our history, and of what Franklin and Whitefield taught us.

ACKNOWLEDGMENTS

Acknowledgments once again. First, to Williamjames Hoffer and N. E. H. Hull, who have never failed to provide essential help with early drafts; next, to Billy G. Smith and Michael Winship, who tried to set me straight on Franklin's Philadelphia and Whitefield's Calvinism, respectively; then, to the kind colleagues and graduate students who attended the University of Georgia Early American History Workshop in Athens, particularly Allan Kulikoff, Laura Davis, and Kylie Horney, I am greatly indebted. As well my thanks to the folks at the McNeil Center for Early American Studies in Philadelphia who attended my session there and commented on chapter two of the manuscript, especially Michelle McDonald, Dallett Hemphill, George Boudreau, and John Murrin. Harry Stout offered an encouraging reading of chapter three. For the press, Bill Pencak gave the manuscript a wonderfully supportive and helpful reading. To Bob Brugger, reader and editor extraordinaire, another round of thanks. Jeremy Horsefield copyedited the manuscript with great care and thoughtfulness. Remaining errors are of course my own responsibility.

NOTES

PROLOGUE: A Momentous Meeting

1. Ralph Waldo Emerson, "The Uses of Great Men," in Emerson, *Addresses and Lectures* (Boston: Houghton Mifflin, 1883), 9.

2. Charles McLean Andrews, *The Colonial Period* (New York: Holt, 1912), 10; Bernard Bailyn, *Atlantic History: Concept and Contours* (New York: Knopf, 2005), 6, 41–44.

3. Aaron Fogelman, *Hopeful Journeys: German Immigration, Settlement, and Political Culture in Colonial America, 1717–1775* (Philadelphia: University of Pennsylvania Press, 1996), 2, table; Christopher Tomlins, *Freedom Bound: Law, Labor, and Civic Identity in Colonizing British America, 1580–1865* (New York: Cambridge University Press, 2010), 43, table 1.6.

4. Jack P. Greene, "Empire and Identity from the Glorious Revolution to the American Revolution," in *The Oxford History of the British Empire*, ed. William Roger Louis et al. (Oxford: Oxford University Press, 2001), 2:208.

5. Bernard Bailyn, *The Peopling of British North America: An Introduction* (New York: Knopf, 1986), 112; Bailyn and John Clive, "England's Cultural Provinces: Scotland and America," *The William and Mary Quarterly*, 3rd ser. 11 (1954), 202–3, 191; David Freeman Hawke, *Everyday Life in Early America* (New York: Harper, 1988), 70–71; James Van Horn Melton, *The Rise of the Public in Enlightenment Europe* (New York: Cambridge University Press, 2001), 81.

6. Claire Kramsch, *Language and Culture* (New York: Oxford University Press, 1998), 5; Richard D. Brown, *Knowledge Is Power: The Diffusion of Information in Early America, 1700–1865* (New York: Oxford University Press, 1989), 43.

7. On the receipts from the selling of Whitefield, see C. William Miller, *Benjamin Franklin's Philadelphia Printing, 1728–1766, A Descriptive Bibliography*, American Philosophical Society Memoirs, vol. 102 (Philadelphia: American Philosophical Society, 1974), 85.

8. Walter Isaacson, *Benjamin Franklin: An American Life* (New York: Simon and Schuster, 2003), 64–72, 94–101.

9. Frank Lambert, *"Pedlar in Divinity": George Whitefield and the Transatlantic Revivals* (Princeton: Princeton University Press, 1994), 52–53, 55.

10. David Waldstreicher, *Runaway America: Benjamin Franklin, Slavery, and the American Revolution* (New York: Hill and Wang, 2004), 115, 120–21; T. H. Breen, *The*

Marketplace of Revolution: How Consumer Politics Shaped American Independence (New York: Oxford University Press, 2004), 35, 54–56.

11. Ian K. Steele, *The English Atlantic, 1675–1740: An Exploration of Communication and Community* (New York: Oxford University Press, 1986), 208.

CHAPTER ONE: A Partnership of Mutual Convenience

1. *Poor Richard's Almanack* (Philadelphia: B. Franklin, 1740), n.p.; John Huxtable Elliott, *Empires of the Atlantic World: Britain and Spain in America 1492–1830* (New Haven: Yale University Press, 2003), 233.

2. *Boston Evening-Post*, September 17, 1739.

3. *Pennsylvania Gazette*, December 22, 1737; *Pennsylvania Gazette*, July 26, 1739. Both squibs were copied from the *Advertiser*; hence, the originals were the work of Seward.

4. *Pennsylvania Gazette*, October 18, 1739; November 8, 1739. Actually, Whitefield's "Bethesda" would be the second orphanage in the colony. The Salzburgers in the Ebenezer, Georgia, settlement had established an orphanage. Clyde E. Buckingham, "Early American Orphanages: Ebenezer and Bethesda," *Social Forces* 26 (1948): 311–12.

5. Arnold A. Dallimore, *George Whitefield* (London: Banner of Truth, 1970), 1:413; Harry S. Stout, *The Divine Dramatist: George Whitefield and the Rise of Modern Evangelism* (Grand Rapids, MI: Eerdmans, 1991), 157–58; John Gillies, *Memoirs of Rev. George Whitefield* (Middleton, CT: Hunt and Noyes, 1837), 42–44.

6. J. A. Leo Lemay, *The Life of Benjamin Franklin, Volume 2: Printer and Publisher, 1730–1747* (Philadelphia: University of Pennsylvania Press, 2006), 422–23; John Thomas Scharf and Thompson Westcott, *History of Philadelphia, 1609–1884* (Everts: Philadelphia, 1884), 2:857, 1,253; Benjamin Franklin, *The Autobiography of Benjamin Franklin*, ed. Leonard W. Labaree et al. (New Haven: Yale University Press, 1964), 76.

7. George Whitefield, *A Continuation of the Reverend Mr. Whitefield's Journal from His Embarking after the Embargo to His Arrival at Savannah in Georgia* (Philadelphia: B. Franklin, 1740), 53, 54; Whitefield, "Walking with God," *The Works of the Reverend George Whitefield* (London, 1772), 5:29; Perry Miller, *Roger Williams: His Contributions to the American Tradition* (New York: Athenaeum, 1962), 37, 45.

8. Whitefield, *Journal from His Embarking*, 54.

9. Ibid., 57, 58.

10. Ibid., 55, 62; Whitefield, *A Continuation of the Reverend Mr. Whitefield's Journal from A Few Days after His Arrival at Georgia to His Second Return Thither from Pennsylvania* (Philadelphia: B. Franklin, 1740), 29, 30, 31; Stout, *Divine Dramatist*, 80–83, 110, 133; Frank Lambert, *"Pedlar in Divinity": George Whitefield and the Transatlantic Revivals* (Princeton: Princeton University Press, 1994), 172–73; Gillies, *Memoirs*, 34, 36–37.

11. Stout, *Divine Dramatist*, 104.

12. Edwards to Whitefield quoted in George Marsden, *Jonathan Edwards: A Life* (New Haven: Yale University Press, 2003), 204; Sara Edwards quoted in Stout, *Divine Dramatist*, 127.

13. Smith and Cole quoted in Sandra M. Gustafson, *Eloquence Is Power: Oratory and Performance in Early America* (Chapel Hill: University of North Carolina Press, 2000), 46–47.

14. Johnson and Hume quoted in Allan Gallay, *The Formation of a Planter Elite: Jonathan Bryan and the Southern Colonial Frontier* (Athens: University of Georgia Press, 2007), 30; Carlton quoted in Jill Lepore, *New York Burning: Liberty, Slavery, and Conspiracy in Eighteenth-Century Manhattan* (New York: Knopf, 2005), 188; Whitefield, *Three Letters from the Reverend Mr. G. Whitefield* (Philadelphia: B. Franklin, 1740), 16.

15. Benjamin Franklin, "Articles of Belief and Acts of Religion," unpublished essay, November 20, 1728, in *Papers of Benjamin Franklin*, ed. Leonard W. Labaree et al. (New Haven: Yale University Press, 1960), 1:101–9.

16. Franklin, "Journal of a Voyage, July 29, 1726," in *Not Your Usual Founding Father: Selected Readings from Benjamin Franklin*, ed. Edmund S. Morgan (New Haven: Yale University Press, 2006), 9.

17. Edmund S. Morgan, *Benjamin Franklin* (New Haven: Yale University Press, 2002), 19–21; Franklin, *Autobiography*, 179.

18. Lambert, "*Pedlar*," 77.

19. Walter Isaacson, *Benjamin Franklin: An American Life* (New York: Simon and Schuster, 2003), 37; Esmond Wright, *Franklin of Philadelphia* (Cambridge, MA: Harvard University Press, 1986), 10, 52; Lambert, "*Pedlar*," 5; Stout, *Divine Dramatist*, 41.

20. Franklin, *Autobiography*, 177; *Pennsylvania Gazette*, November 8, 1739; November 15, 1739.

21. Franklin, *Autobiography*, 167–68; [Franklin], "Observations on the Proceedings against Mr. Hemphill," 1735, *Papers of Franklin*, 2:45, 60; LeMay, *Life of Franklin*, 2:242.

22. Franklin, *Autobiography*, 175–76.

23. Ibid., 175–76, 177.

24. Whitefield, *Journal from His Embarking*, 61; *Pennsylvania Gazette*, December 6, 1739.

25. Whitefield, *Journal from His Embarking*, 63; *Pennsylvania Gazette*, November 15, 1739; November 22, 1739; November 29, 1739; Franklin, writing as Obadiah Plainman, *Pennsylvania Gazette*, May 15, 1740.

26. *Pennsylvania Gazette*, December 6, 1739; Stout, *Divide Dramatist*, 110, 120; Lemay, *Life of Franklin*, 2:424; Hubertis Maurice Cummings, *Richard Peters: Provincial Secretary and Cleric, 1704–1776* (Philadelphia: University of Pennsylvania Press, 1944), 44.

27. Whitefield to Franklin, November 26, 1740, *A Select Collection of Letters of the Late George Whitefield, M.A. . . .* (London, 1772), 1:226.

CHAPTER TWO: Franklin Becomes a Printer and Whitefield Becomes a Preacher

1. Benjamin Franklin, *The Autobiography of Benjamin Franklin*, ed. Leonard W. Labaree et al. (New Haven: Yale University Press, 1964), 180; but Franklin did not always play fair with facts; Gordon S. Wood, *The Americanization of Benjamin Franklin* (New York: Penguin, 2004), 13–14.

2. Silence Dogood, *The New-England Courant*, April 16, 1722.

3. Ibid., May 14, 1722.

4. Carl Bridenbaugh, *Cities in the Wilderness: The First Century of Urban Life in America 1625–1742* (New York: Knopf, 1956), 83, 217.

5. On social capital, see Jack P. Greene, "Social and Cultural Capital in British North America: A Case Study," *Journal of Interdisciplinary History* 29 (1999): 491–509.

6. Franklin, *Autobiography*, 75.

7. J. H. Powell, *Bring Out Your Dead: The Great Plague of Yellow Fever in Philadelphia in 1793* (Philadelphia: University of Pennsylvania Press, 1993), 1–2; Billy G. Smith, *Life in Early Philadelphia: Documents from the Revolutionary and Early National Periods* (State College: Pennsylvania State University Press, 1995), 225–29; J. A. Leo Lemay, *The Life of Benjamin Franklin, Volume 2: Printer and Publisher, 1730–1747* (Philadelphia: University of Pennsylvania Press, 2006), 2:24.

8. Alexander Hamilton, "The Itinerarium of Dr. Alexander Hamilton," in *Colonial American Travel Narratives*, ed. Wendy Martin (New York: Penguin, 1994), 189, 192.

9. Ibid., 197.

10. Frederick B. Tolles, *Meeting House and Counting House: The Quaker Merchants of Colonial Philadelphia* (Chapel Hill: University of North Carolina Press, 1948), 48, 50; Edwin B. Bronner, "Village into Town 1701–1744," in *Philadelphia: A 300-Year History*, ed. Russell F. Weigley (New York: Norton, 1982), 33–37; Peter Kalm, "Travels into North America" [1753], reprinted in Oscar Handlin, ed., *This Was America* (New York: Harper, 1949), 22.

11. Arthur Jensen, *The Maritime Commerce of Colonial Philadelphia* (Madison: Wisconsin Historical Society, 1963), 135.

12. Hamilton, "Itinerarium," 191, 192, 319.

13. Bridenbaugh, *Cities in the Wilderness*, 357–58, 394–95; Billy G. Smith, "The Vicissitudes of Fortune: The Career of Laboring Men in Philadelphia, 1750–1800," in *Work and Labor in Early America*, ed. Stephen Innes (Chapel Hill: University of North Carolina Press, 1988), 221–51; Gary B. Nash, *The Urban Crucible: Social Change, Political Consciousness, and the Origins of the American Revolution* (Cambridge, MA: Harvard University Press, 1979), 74.

14. Lemay, *Life of Franklin*, 2:278; Gary B. Nash, *Forging Freedom: The Formation of Philadelphia's Black Community, 1720–1840* (Cambridge, MA: Harvard University Press, 1988), 9–14.

15. Hamilton, "Itinerarium," 320; Patricia U. Bonomi, *Under the Cope of Heaven:*

Religion, Society, and Politics in Colonial America (New York: Oxford University Press, 1986), 90.

16. Esmond Wright, *Franklin of Philadelphia* (Cambridge, MA: Harvard University Press, 1986), 29; H. W. Brands, *The First American: The Life and Times of Benjamin Franklin* (New York: Doubleday, 2000), 40; Walter Isaacson, *Benjamin Franklin: An American Life* (New York: Simon and Schuster, 2003), 37, 39; J. A. Leo Lemay, *The Life of Benjamin Franklin, Volume 1: Journalist, 1706–1730* (Philadelphia: University of Pennsylvania Press, 2005), 206.

17. Gordon S. Wood, *The Radicalism of the American Revolution* (New York: Knopf, 1992), 57, 88.

18. Franklin, "A Dissertation on Liberty and Necessity, Pleasure and Pain" (1725), in *Papers of Benjamin Franklin*, ed. Leonard W. Labaree et al. (New Haven: Yale University Press, 1960), 1:507.

19. Franklin to Whitefield, June 19, 1764, *Papers of Franklin*, 11:231.

20. Thomas Doerflinger, *A Vigorous Spirit of Enterprise: Merchants and Economic Development in Revolutionary Philadelphia* (Chapel Hill: University of North Carolina Press, 1986), 53.

21. Adrian Johns, *Piracy: The Intellectual Property Wars from Gutenberg to Gates* (Chicago: University of Chicago Press, 2009), 150.

22. Quotations in Brands, *First American*, 121; Isaacson, *Benjamin Franklin*, 95–96; Franklin, *Autobiography*, 163–64.

23. Silence Dogood, *The New-England Courant*, October 8, 1722; Franklin, *Poor Richard's Almanack* (Philadelphia: B. Franklin, 1743), n.p.

24. Richard Sennett, *The Fall of Public Man* (New York: Norton, 1992), 284, 21; Tolles, *Meeting House and Counting House*, 247.

25. George Whitefield, *A Short Account of God's Dealings with the Reverend George Whitefield, Written by Himself* (London, 1740), preface.

26. Ibid., 13.

27. Ibid., 13; Whitefield, "Walking with God," in *Works of George Whitefield* (London, 1771), 5:27.

28. Harry S. Stout, *The Divine Dramatist: George Whitefield and the Rise of Modern Evangelism* (Grand Rapids, MI: Eerdmans, 1991), 5, 11, 12.

29. Ronald H. Quilici, "Turmoil in a City and an Empire: Bristol's Factions, 1700–1775" (Ph.D. diss., University of New Hampshire, 1976), 140; C. M. MacInnes, *A Gateway of Empire* (New York: Kelly, [1939] 1968), 180, 217, 241, 261.

30. Brian S. Smith and Elizabeth Ralph, *A History of Bristol and Gloucestershire* (Chichester, UK: Phillimore, 1996), 94–96; Quilici, "Turmoil in a City and an Empire," 28, 73, 203.

31. Stout, *Divine Dramatist*, 19.

32. Jonathan Edwards, "Sinners in the Hands of an Angry God," July 8, 1741, in *Jonathan Edwards's "Sinners in the Hands of an Angry God": A Casebook*, ed. Wilson H. Kimnach, Caleb J. D. Maskell, and Kenneth P. Minkema (New Haven: Yale University Press, 2010), 34; George Whitefield, "Law Gospelized, or An Address to All

Christians," in *Works of the Reverend George Whitefield*, 6 vols. (London: Dilly, 1771–1772), 4:400.

33. Whitefield, "The Potter and the Clay," *Works of Whitefield*, 5:214; Stout, *Divine Dramatist*, 37.

34. Whitefield to Franklin, June 23, 1747, *Papers of Franklin*, 3:143.

35. John Lawson, *A New Voyage to Carolina* (London, 1709), 1, 3.

36. Frank Lambert, *"Pedlar in Divinity": George Whitefield and the Transatlantic Revivals* (Princeton: Princeton University Press, 1994), 55.

37. Clare Brant and Susan E. Whyman, eds., "Introduction," *Walking the Streets of Eighteenth-Century London: John Gay's* Trivia *(1716)* (New York: Oxford University Press, 2007), 3; Rosemary Sweet and Penelope Lane, *Women and Urban Life in Eighteenth-Century London: Prostitution and Control in the Metropolis, 1730–1830* (London: Longman, 1999), 52.

38. Peter Earle, *The Making of the English Middle Class: Business, Society, and Family Life in London, 1663–1730* (Berkeley: University of California Press, 1989), 3, 6, 7, 8; Joanna Innes, *Inferior Politics: Social Problems and Social Policies in Eighteenth-Century Britain* (New York: Oxford University Press, 2009), 280; T. S. Ashton, *An Economic History of England: The Eighteenth-Century* (London: Taylor and Francis, 2006), 20, 162.

39. Fielding quoted in Earle, *Middle Class*, 11.

40. Lambert, *"Pedlar,"* 62, 65.

41. Whitefield to John Wesley, quoted in John Gillies, *Memoirs of Rev. George Whitefield* (Middleton, CT: Hunt and Noyes, 1837), 56; John Hurst, *John Wesley the Methodist: A Plain Account of His Life and Work* (London: Eaton and Mains, 1903), 154–55.

42. Lambert, *"Pedlar,"* 63.

CHAPTER THREE: Whitefield's Messages of Hope

1. The publications are listed with bibliographical notes in *Papers of Benjamin Franklin*, ed. Leonard W. Labaree et al. (New Haven: Yale University Press, 1960), 2:243n2.

2. Frank Lambert, *"Pedlar in Divinity": George Whitefield and the Transatlantic Revivals* (Princeton: Princeton University Press, 1994), 77.

3. George Whitefield, *Three Letters* (Philadelphia: B. Franklin, 1740), originally published in issues of the *Pennsylvania Gazette* from April 10 to May 1, 1740. See J. A. Leo Lemay, *The Life of Benjamin Franklin, Volume 2: Printer and Publisher, 1730–1747* (Philadelphia: University of Pennsylvania Press, 2006), 424–25.

4. *Pennsylvania Gazette*, December 13, 1739.

5. Lambert, *"Pedlar,"* 70–72.

6. Stephen Foster, *The Long Argument: English Puritanism and the Shaping of New England Culture, 1570–1700* (Chapel Hill: University of North Carolina Press, 1991), 294; Whitefield, "Walking with God," in *Works of the Reverend George Whitefield*, 6

vols. (London: Dilly, 1771–1772), 5:30; Bruce C. Daniels, *Puritans at Play: Leisure and Recreation in Colonial New England* (New York: Macmillan, 1996), 32.

7. Here and after, George Whitefield, "Abraham's Offering Up His Son Isaac," in John Gillies, *Memoirs of Rev. George Whitefield* (Middletown, CT: Hunt and Noyes, 1837), 339–50.

8. Michael Winship, *Making Heretics: Militant Protestantism and Free Grace in Massachusetts, 1636–1641* (Princeton: Princeton University Press, 2000), 84, 182.

9. John Wesley, "Predestination Calmly Considered" [1752], *The Miscellaneous Works of the Reverend John Wesley* (New York: Harper, 1828), 2:377, 393.

10. John Winthrop, lay sermon on the Arbella, at sea, 1630, in *The Puritans*, ed. Perry Miller and Thomas H. Johnson (New York: Harper, 1938), 1:195.

11. Whitefield to Lady G___, February 22, 1749, *Works of Whitefield*, 2:234–35; Carolyn Haynes, *Divine Destiny: Gender and Race in Nineteenth-Century Protestantism* (Oxford: University Press of Mississippi, 1999), 17.

12. Whitefield to G___P___, September 30, 1752, *Works of Whitefield*, 2:447. The controversy is explained in Winship, *Making Heretics*, 69–71.

13. Richard T. Eldridge, *An Introduction to the Philosophy of Art* (New York: Cambridge University Press, 2003), 198–200; on Aristotle and Whitefield, see, e.g., Whitefield, "Britain's Mercies and Britain's Duties" [1746], *Sermons on Important Subjects by the Reverend George Whitefield*, ed. S. Drew (London: Fisher, 1828), 96.

14. Whitefield, "Marks of a True Conversion," *Sermons on Important Subjects*, 271, 277, 275.

15. E. A. Wrigley, *Progress, Poverty, and Population* (Cambridge: Cambridge University Press, 2004), 417; Wrigley, et al., *English Population History from Family Reconstitution 1580–1837* (Cambridge: Cambridge University Press, 2005), 206; J. H. Hays, *The Burdens of Disease: Epidemics and Human Response in Western History* (New Brunswick, NJ: Rutgers University Press, 2009), 105–6.

16. Whitefield, "A Second Letter to the Right Reverend, the Bishop of London, 1744," *Works of Whitefield* 4:163; Whitefield, "All Men's Place," *Eighteen Sermons* (Newburyport, MA, 1820), 235.

17. Whitefield, Journal, October 18–19, 1740, quoted in Arnold A. Dallimore, *George Whitefield* (London: Banner of Truth, 1970), 537.

18. Whitefield to Mr. ___, June 8, 1753, *A Selection of Letters*, 2:16; Whitefield to Mr. D___, April 27, 1753, ibid., 2:11; Whitefield to Mr. S___, May 27, 1753, ibid., 2:13.

19. Whitefield to John Wesley, December 24, 1740, *Works of Whitefield*, 4:70, 72.

20. Whitefield, "The Temptation of Christ," *Works of Whitefield*, 5:270; "Walk with God," *Works of Whitefield*, 5:36; Whitefield to Franklin, February 26, 1750, *Papers of Franklin*, 3:467.

CHAPTER FOUR: Franklin's Essays on Improvement

1. Franklin, "Busy Body No. 1," *American Weekly Mercury*, February 4, 1729.

2. Franklin, "The Printer to the Reader," *Pennsylvania Gazette*, October 2, 1729.

3. Franklin, "A Witch Trial at Mount Holly," *Pennsylvania Gazette*, October 22, 1730.

4. On colonial taverns, see generally Sharon V. Salinger, *Taverns and Drinking in Early America* (Baltimore: Johns Hopkins University Press, 2004).

5. Franklin, "The Drinker's Dictionary," *Pennsylvania Gazette*, January 13, 1737.

6. Benjamin Franklin, *A Proposal for Promoting Useful Knowledge among the British Plantations in America* (Philadelphia: B. Franklin, 1743).

7. John Brewer, *The Sinews of Power: War, Money, and the English State, 1688–1783* (Cambridge, MA: Harvard University Press, 1990), 15, 67, 69, 225.

8. Benjamin Franklin, *The Autobiography of Benjamin Franklin*, ed. Leonard W. Labaree et al. (New Haven: Yale University Press, 1964), 182–83; John W. Jordan, *Colonial and Revolutionary Families of Pennsylvania* (New York, 1911), 2:88; Penn quoted in Lorraine Smith Pangle, *The Political Philosophy of Benjamin Franklin* (Baltimore: Johns Hopkins University Press, 2007), 115. On Norris's copy, see www.librarycompany.org/bfwriter/plain.htm.

9. Here and after Benjamin Franklin, *Plain Truth: Serious Considerations on the Present State of the City of Philadelphia and Province of Pennsylvania* (Philadelphia: B. Franklin, 1747), 3–10.

10. Franklin quoted in James Merrell, *Into the American Woods: Negotiations on the Pennsylvania Frontier* (New York: Norton, 2000), 258.

11. Ian Steele, *Warpaths: Invasions of North America* (New York: Oxford University Press, 1995), 169–201.

12. Daniel Richter, *Ordeal of the Longhouse: The Peoples of the Iroquois League in the Era of European Colonization* (Chapel Hill: University of North Carolina Press, 1992), 271.

13. Merrell, *American Woods*, 38, 67; Fred Anderson, *Crucible of War: The Seven Years' War and the Fate of Empire in British North America, 1754–1766* (New York: Knopf, 2000), 28–29.

14. Gordon S. Wood, *The Radicalism of the American Revolution* (New York: Knopf, 1992), 86.

CHAPTER FIVE: A Great Awakening, the Enlightenment, and the Crisis of Provincialism

1. T. H. Breen, *The Marketplace of Revolution: How Consumer Politics Shaped American Independence* (New York: Oxford University Press, 2004), 104; Henry May, *The Enlightenment in America* (New York: Oxford University Press, 1976), 42, 48, 85.

2. Whitefield, "great and general awakening" quotation in Joseph Tracy, *The Great Awakening* (Boston: Tappan and Dennett, 1842), 102; "son of thunder" quotation in Alan Craig Houston, *Benjamin Franklin and the Politics of Improvement* (New Haven: Yale University Press, 2008), 71.

3. Jonathan Edwards, "A Faithful Narrative," in *A Jonathan Edwards Reader*, ed. John Smith and Harry S. Stout (New Haven: Yale University Press, 2003), 58, 59, 60.

4. Samuel Hopkins, "System of Doctrines," *Works of Samuel Hopkins* (Boston: Doctrinal Tract and Book Society, 1852), 2:3; John William Fletcher, "Zelotes and Honestus Reconciled, or The Third Part of An Equal Check to Pharisaism and Antinomianism," *The Works of the Reverend John Fletcher* (Philadelphia: Lane and Scott, 1851), 2:131, 235; Stephen Foster, *The Long Argument: English Puritanism and the Shaping of New England Culture, 1570–1700* (Chapel Hill: University of North Carolina Press, 1991), 300.

5. Foster, *Long Argument*, 297.

6. Prince quoted in Arnold A. Dallimore, *George Whitefield* (London: Banner of Truth, 1970), 141.

7. Alexander Garden, *Regeneration and the Testimony of the Spirit* [1740], reprinted in Alan Heimert and Perry Miller, eds., *The Great Awakening* (Indianapolis: Bobbs-Merrill, 1967), 51; Charles Chauncy, *Enthusiasm Described and Cautioned Against* [1742], in ibid., 229.

8. Whitefield, Journal for May 5, 1738, quoted in Dallimore, *Whitefield*, 51.

9. Whitefield quoted in Richard Bushman, ed., *The Great Awakening: Documents on the Revival of Religion, 1740–1745* (Chapel Hill: University of North Carolina Press, 1989), 5.

10. Davenport, *Reverend Mr. Davenport's Song* (Boston, 1742), 2; *Boston Weekly Post-Boy*, March 28, 1743, quoted in Bushman, ed., *Great Awakening*, 52; *Christian History* 2:406, 408; C. C. Goen, *Revivalism and Separatism in New England, 1740–1800* (New Haven, 1962), 24–25; Joshua Hempstead, "Diary," March 27, 1743, in J. M. Bumsted, ed., *The Great Awakening: The Beginnings of Evangelical Pietism in America* (Waltham, MA: Blaisdell, 1969), 89.

11. Davenport, *The Reverend Mr. Davenport's Confession and Retraction* (Boston, 1744), 5; Davenport, *A Letter from Mr. James Davenport to Mr. Jonathan Barber* (Philadelphia, 1744), 4, 6–7.

12. Whitefield, Journal, November 27, 1744, reproduced in Bushman, ed., *Great Awakening*, 65; Harry S. Stout, *The Divine Dramatist: George Whitefield and the Rise of Modern Evangelism* (Grand Rapids, MI: Eerdmans, 1991), 174–75, 178–79, 192; Joseph A. Conforti, *Jonathan Edwards, Religious Tradition, and American Culture* (Chapel Hill: University of North Carolina Press, 1995), 199n5.

13. George Whitefield to Benjamin Franklin, April 16, 1746, *Papers of Benjamin Franklin*, ed. Leonard W. Labaree et al. (New Haven: Yale University Press, 1960), 3:71–74. The continuing financial travail of the orphanage is recounted in Edward J. Cashin, *Beloved Bethesda: A History of George Whitefield's Home for Boys, 1740–1900* (Macon, GA: Mercer University Press, 2001), 37ff.

14. Stout, *Divine Dramatist*, 199; Christine Leigh Heyrman, *Southern Cross: The Beginnings of the Bible Belt* (New York: Knopf, 1997), 255–56.

15. Stout, *Divine Dramatist*, 176; Frank Lambert, *"Pedlar in Divinity": George Whitefield and the Transatlantic Revivals* (Princeton: Princeton University Press, 1994), 201–2; Colman quoted in Lambert, *"Pedlar,"* 199.

16. Stout, *Divine Dramatist*, 172–73, 213–15; *The Life and Times of Selina Countess*

of Huntington: By a Member of the Houses of Shirley and Hastings (London: Painter, 1841), 1:184.

17. Benjamin Franklin to George Whitefield, July 6, 1749, *Evangelical Magazine* 11 (1803): 28.

18. Peter Gay, *The Enlightenment: The Science of Freedom* (New York: Norton, 1996), 3; Isaac Kramnick, "Introduction," in Kramnick, ed., *The Portable Enlightenment Reader* (New York: Penguin, 1995), xxi, xix.

19. John Richetti, "Introduction," *The Cambridge Companion to the Eighteenth-Century Novel* (Cambridge: Cambridge University Press, 1996), 2–3; Eve Rachele Sanders, *Gender and Literacy on Stage in Early Modern England* (Cambridge: Cambridge University Press, 1998), 143.

20. Barbara Shapiro, *A Culture of Fact: England, 1550–1720* (Ithaca: Cornell University Press, 2000), 35, 37; Jeremy Black, *The English Press in the Eighteenth-Century* (London: Taylor and Francis, 1989), 9, 13.

21. Gordon S. Wood, *The Americanization of Benjamin Franklin* (New York: Penguin, 2004), 56–57.

22. I. Bernard Cohen, *The Revolution in Science* (Cambridge, MA: Harvard University Press, 1987), 174–75.

23. Benjamin Franklin, *The Autobiography of Benjamin Franklin*, ed. Leonard W. Labaree et al. (New Haven: Yale University Press, 1964), 196.

24. Franklin to Peter Collinson, March 28, 1747, *Papers of Franklin*, 3:115.

25. I. Bernard Cohen, *Benjamin Franklin's Science* (Cambridge, MA: Harvard University Press, 1990), 13–16; Sir Thomas Sprat, *The History of the Institution, Design and Progress of the Royal Society of London . . .* , 3rd ed. (London, 1722), 346; Wood, *Americanization*, 65.

26. Franklin, *Experiments and Observations on Electricity, Made at Philadelphia in America*, 4th ed. (London: Henry, [1751] 1769), 84; Joyce Chaplin, *The First Scientific American: Benjamin Franklin and the Pursuit of Genius* (New York: Basic Books, 2006), 110.

27. Wood, *Americanization*, 64.

28. Franklin, *Experiments*, 8.

29. Joyce Chaplin, *Scientific American*, 7, 108; Walter Isaacson, *Benjamin Franklin: An American Life* (New York: Simon and Schuster, 2003), 134; Paul Hyland, Olga Gomez, and Francesca Greensides, *The Enlightenment: A Sourcebook and Reader* (London: Routledge, 2003), 124.

30. Isaacson, *Benjamin Franklin*, 129, 142–45.

31. H. W. Brands, *The First American: The Life and Times of Benjamin Franklin* (New York: Doubleday, 2000), 166.

32. Samuel Y. Edgerton Jr., "Supplement: The Franklin Stove," in Cohen, *Franklin's Science*, 199–211; Franklin quoted in Cohen, *Franklin's Science*, 165.

33. Franklin, *Autobiography*, 196; Brands, *First American*, 181–86.

34. Fred Anderson and Andrew W. Cayton, *The Dominion of War: Empire and Liberty in North America, 1500–2000* (New York: Viking, 2004), 122.

35. Here and after [Franklin], "Short Hints towards a Scheme for a General Union of the British Colonies on the Continent," *Papers of Franklin*, 5:357.

36. Alan Rogers, *Empire and Liberty: American Resistance to British Authority, 1755–1763* (Berkeley: University of California Press, 1974), 11–18.

37. Adam Smith, *Inquiry into the Wealth of Nations* (New York: Collier, [1776] 1902), 2:394; Wood, *Americanization*, 113.

38. The account, with full citations, appears in my *Sensory Worlds in Early America* (Baltimore: Johns Hopkins University Press, 2003), 207–17.

39. Brands, *First American*, 367, quotation on 368.

40. Whitefield quoted in Stout, *Divine Dramatist*, 263.

41. Franklin to the *Gazetteer and New Daily Advertiser*, December 28, 1765.

42. Tal Golan, *Laws of Men and Laws of Nature: The History of Scientific Expert Testimony in England and America* (Cambridge, MA: Harvard University Press, 2004), 16.

43. Kirstin Olsen, *Daily Life in Eighteenth-Century England* (Santa Barbara, CA: Greenwood, 1999), 6–7.

44. Hannah Barker, *Newspapers, Politics, and Public Opinion in Late Eighteenth-Century England* (Oxford: Oxford University Press, 1982), 17, 19; Jeremy Black, *The English Press in the Eighteenth Century* (London: Taylor and Francis, 1987), 129.

45. See, e.g., Richard Beeman, *The Varieties of Political Experience in Eighteenth-Century America* (Philadelphia: University of Pennsylvania Press, 2004), 52ff.

46. Here and after *The Examination of Doctor Benjamin Franklin, before an August Assembly, relating to the Repeal of the Stamp Act, &c.* (Philadelphia: Hall and Sellers, 1766); the text here is from *Papers of Franklin*, 13:124–58.

47. Franklin to Whitefield, July 2, 1756, *Papers of Franklin*, 6:468; David Morgan, *The Devious Dr. Franklin, Colonial Agent: Benjamin Franklin's Years in London* (Macon, GA: Mercer University Press, 1999), 76.

48. Richard L. Bushman, *King and People in Provincial Massachusetts* (Chapel Hill: University of North Carolina Press, 1992), 5, 23.

49. This text derived from Peter Charles Hoffer, *Law and People in Colonial America*, rev. ed. (Baltimore: Johns Hopkins University Press, 1998), 151–52.

50. Washington quoted in Michael Zakim, "Sartorial Ideologies: From Homespun to Ready-Made," *American Historical Review* 106 (2001): 1556.

51. Morgan, *Devious Dr. Franklin*, 159; Franklin to Whitefield, before September 2, 1769, *Papers of Franklin*, 16:92.

52. Franklin to Whitefield, before September 2, 1769, *Papers of Franklin*, 16:192.

53. Whitefield to Franklin, January 21, 1768, *Papers of Franklin*, 15:28; Stout, *Divine Dramatist*, 269–80.

EPILOGUE: The Birth of the Modern World

1. David Waldstreicher, *Runaway America: Benjamin Franklin, Slavery, and the American Revolution* (New York: Hill and Wang, 2004), 199; Edmund S. Morgan, *Benjamin Franklin* (New Haven: Yale University Press, 2002), 97, 240.

2. Christopher Lasch, *The True and Only Heaven: Progress and Its Critics* (New York: Norton, 1991), 14; Nathan Hatch, Mark A. Noll, and George Marsden, *The Search for a Christian America* (Colorado Springs: Helmers and Howard, 1989), 15.

3. Nathan Hatch, *The Democratization of American Christianity* (New Haven: Yale University Press, 1989), 225.

4. J. Richard Olivas, "Partial Revival: The Limitations of the Great Awakening in Boston, 1740–1742," in *Inequality in Early America*, ed. Carla Gardina Pestana and Sharon V. Salinger (Hanover, NH: University Press of New England, 1999), 67, 75–76.

5. Esmond Wright, *Franklin of Philadelphia* (Cambridge, MA: Harvard University Press, 1986), 358; Michael Zuckerman, *Almost Chosen People: Oblique Biographies in the American Grain* (Berkeley: University of California Press, 1993), 146.

6. Carl Van Doren, *Benjamin Franklin* (New York: Viking, 1938), 779.

7. Richard D. Brown, *Modernization: The Transformation of American Life, 1600–1865* (New York: Hill and Wang, 1976), 53, 76; Franklin quoted in Van Doren, *Franklin*, 312.

8. James W. Ceaser, *Reconstructing America: The Symbol of America in Modern Thought* (New Haven: Yale University Press, 2000), 44, 166.

9. Joyce Appleby, *Inheriting the Revolution: The First Generation of Americans* (Cambridge, MA: Harvard University Press, 2000), 117; Charles E. Hambrick-Stowe, *Charles G. Finney and the Spirit of American Evangelicism* (Grand Rapids, MI: Eerdmans, 1996), 243; Barry Hankins, *The Second Great Awakening and the Transcendentalists* (Santa Barbara, CA: Greenwood, 2004), 44.

10. Jerome Dean Mahaffey, *Preaching Politics: The Religious Rhetoric of George Whitefield and the Founding of a New Nation* (Waco, TX: Baylor University Press), 200; Hatch, *Democratization*, 127.

11. I. Bernard Cohen, *Benjamin Franklin's Science* (Cambridge, MA: Harvard University Press, 1990), 6–7.

12. Stephen Toulmin, *Cosmopolis: The Hidden Agenda of Modernity* (Chicago: University of Chicago Press, 1992), 3.

ESSAY ON SOURCES

The present volume grows out of an earlier work on the sensory world of the American colonists, Peter Charles Hoffer, *Sensory Worlds in Early America* (Baltimore: Johns Hopkins University Press, 2003), in which Franklin and Whitefield appeared. It was clear that both men fashioned themselves to fit perceptions of their audience and were masters of communication. I did not argue that they were the first moderns, however. As a founding coeditor of the Witness to History series, a project designed for classroom use, I began to look at authors' proposals in light of their relevance to modern students' interests. This brought me back to Franklin and Whitefield and enabled me to see them in a new light.

The following bibliographical essay is hardly exhaustive. Such a project would have entailed as many pages as the preceding essay and would be out of date by the time the book was published. Works on collateral subjects, for example, London and Philadelphia in the eighteenth century, contemporary oratory, the history of science, and the Atlantic World, are cited in the notes. The bibliographical essay, like the book itself, is selective and focuses on the two men's connection.

Franklin is not without his critics, but far more common are admiring biographers. The first of the modern biographies was Carl Van Doren's *Benjamin Franklin* (New York: Viking, 1938), a Pulitzer Prize–winning giant (over 830 pages) including a good selection of Franklin's own writing. Simon and Schuster reissued the work in 2002. Van Doren was a professor of American literature at Columbia University, and he viewed Franklin as one of the founding fathers of American letters as well as the new republic. He lionized his subject. "In any age, Franklin would have been great" (782).

A selection of more recent works shows that passing time has not dimmed Franklin's luster. H. W. Brands is an academic biographer whose productivity is simply amazing. His *The First American: The Life and Times of Benjamin Franklin* (New York: Doubleday, 2000) is immensely detailed and balanced, that is, neither exaggerating nor diminishing Franklin's contributions and his qualities. Though "a life as full as Franklin's could not be captured in a phrase—or a volume" (716), Brands appreciates Franklin's legacy, particularly in his later-life contributions to the creation of the American republic.

Edwin S. Gaustad's compact biography *Benjamin Franklin* (New York: Oxford University Press, 2006) has the virtues of succinctness and a fine eye for the telling short quotation, beginning with a comparison of Franklin to Mark Twain (both satirists) and ending by extolling him as an American icon, the author's solemnity replacing the subject's wit. It is part of a series of "Lives and Legacies" from Oxford, celebrating the lives and works of American men of letters.

Walter Isaacson's *Benjamin Franklin: A Life* (New York: Simon and Schuster, 2004) is a big, breezy, highly anecdotal account of the man by a highly successful journalist who I conclude sees much of himself in Franklin. Apparently, they are both pragmatic, keeping an eye on the sales of their publications. The work is organized chronologically, focusing on the man, his family, his character, and his achievements, all "middle class" or "Main Street" values and virtues (493, 492).

J. A. Leo Lemay's *The Life of Benjamin Franklin* (Philadelphia: University of Pennsylvania Press, 2006), in 3 vols., is compendious, highly complimentary to its subject, and a somewhat disorderly account of Franklin from his birth to 1757. Lemay passed away before he could complete the project. All three volumes follow tangents, add collateral material, and wander all over the first two-thirds of Franklin's life. It was obviously a work of love, however. Lemay has also edited selections from Franklin's writings, fittingly, as Lemay was "a literary historian" first and foremost (*The Life of Benjamin Franklin*, Vol. 3, *Soldier, Scientist, and Politician, 1748–1757* [2008], 589).

Edmund S. Morgan's *Benjamin Franklin* (New Haven: Yale University Press, 2003) is a relatively short (350 pages) musing on Franklin and his times. Morgan is one of his generation's greatest social historians and sees behind the "affability and wit" to a man whose "wisdom about himself . . . comes only to the great of heart" (314). Morgan, like Lemay, edited a volume of Franklin's writings.

Gordon S. Wood, one of the premier colonial and early national intellectual historians of the second half of the twentieth century, has a distinctive take on Franklin in *The Americanization of Benjamin Franklin* (New York: Penguin, 2004). In a sense, he is out to debunk the heroic mythology and finds instead a striver seeking to become a gentleman (which he does), who then tries to become an imperial power broker (in which quest he embarrassingly fails), and almost by default becomes a patriot and founding father. It is a tour de force of revisionist biography that will by its own intellectual authority become part of the Franklin canon. But Wood is not unappreciative of Franklin's achievements. What is more, Wood sees in Franklin's story (though not necessarily the man himself) the basis for the enduring myths of American exceptionalism; out of the "repeated messages of striving and success" of later writings about Franklin, Americans were "able to construct an enduring sense of American nationhood—a sense of America as the land of enterprise and opportunity, as the place where anybody who works hard can make it" (243, 244). Though Franklin lived in a time of slavery and inequality, his image came to symbolize democracy and equality.

Esmond Wright's *Franklin of Philadelphia* (Cambridge, MA: Harvard University Press, 1986) is a cool but not distant assessment of Franklin by a leading English historian. It finds the many paradoxes in the subject's life and concludes that Franklin remains elusive because he wanted it that way. In all these paradoxes, he was "'the new man' of the eighteenth-century dream," the Yankee tradesman who is best known for what he does, not what he believes (360).

All of these biographies are largely based on the same set of primary sources. Jared Sparks was the first editor of Franklin's letters and essays, in 1836. Sparks, a minister by training and a moralist by inclination, was a documentary editor and biographer as well. He had no hesitation about leaving out parts of letters he considered injurious to the reputation of the authors (or changing a word here and there for the same purpose).

The gold standard for Franklin is the Yale University–American Philosophical Society edition of the *Papers of Benjamin Franklin* (36 vols.; New Haven: Yale University Press, 1959–1999), first edited by Leonard Labaree and then volumes 15–26 edited by W. B. Willcox, volume 27 by Claude A. Lopez, volumes 28–35 by Barbara B. Oberg, and volume 36 by Ellen R. Cohn. These are models of their kind, with headnotes and citations that explain terms, fill in details, and otherwise illuminate an illuminating life. The entire thirty-six-volume edition (less the editors' extensive annotation) is online in very usable format at www.franklinpapers.org/franklin/framedVolumes.jsp.

A separate volume in the series, *The Autobiography of Benjamin Franklin* (New Haven: Yale University Press, 1964), has a history of its own. He began the memoir in 1771, while he lived in England. It was a work of memory in leisure time. It was not an interior work, in the sense that Franklin did not examine his inner thoughts, but then, much of Enlightenment literature was similar, in the sense of narrating events and motives rather than probing the dark side of the subconscious. He finished Part I but was not able to return to the project (or obtain the original manuscript) until 1784, and then his labors for the new United States of America precluded sustained work on the memoir until 1786. Parts III and IV were written in some haste in 1788, for Franklin's powers of concentration were already deserting him. He died in 1790, having completed the memoir to the year 1758. Rival editions of the first part of the memoir appeared in 1791 and 1793, the first based on a copy that Franklin's grandson, Benjamin Franklin Bache, made, the second the work of an English publishing house. Temple Franklin, Benjamin's literary executor and heir, did not publish a version, though he had the papers from the time of his father's death until 1817. In 1867, when a copy of the original surfaced, it was discovered that Temple Franklin had changed many passages. John Bigelow edited the newly discovered version in 1868, but this too had errors. Max Farrand, a leading historical editor, corrected many of these in a 1949 edition, providing all four previous versions. The Yale edition reproduced the first of these, closest in time and style to Franklin's own writing. William Pencak relates that Paul Zall and Lemay found "the definitive" copy of the original in the Huntington Library, in California.

Whitefield's life has not generated the continuing interest one finds in Franklin. The late eighteenth and early nineteenth centuries held him in higher regard. Major biographies appeared then, two of them including copious selections from his sermons. John Gillies's *Memoirs of Rev. George Whitefield* (Middletown, CT: Hunt and

Noyes, 1837) reprinted many of the sermons. The other was A. S. Billingsley's *The Life of George Whitefield, "Prince of Pulpit Orators," with Specimens of His Sermons* (n.p., 1878). Other works were subtitled "matchless soul winner," "lessons from the life," and the like. They abound.

Among the more recent works are John Pollock, *George Whitefield and the Great Awakening* (Garden City, NY: Doubleday, 1972); Arnold A. Dallimore, *George Whitefield: The Life and Times of the Great Evangelist of the Eighteenth-Century Revival* (2 vols.; Westchester, IL: Cornerstone Books, 1979); Harry S. Stout, *The Divine Dramatist: George Whitefield and the Rise of Modern Evangelicalism* (Grand Rapids, MI: Eerdmans, 1991); and Frank Lambert, *"Pedlar in Divinity": George Whitefield and the Transatlantic Revivals, 1737–1770* (Princeton: Princeton University Press, 1994). Stout emphasized Whitefield's background in theater and traced how it influenced his later career. Lambert's treatment is more topical, focusing on the commercial side of Whitefield's ministry.

Other special topics are the focus of Edward J. Cashin, *Beloved Bethesda: A History of George Whitefield's Home for Boys, 1740–1900* (Macon, GA: Mercer University Press, 2001); Jerome Dean Mahaffey, *Preaching Politics: The Religious Rhetoric of George Whitefield and the Founding of a New Nation* (Waco, TX: Baylor University Press, 2007); and James L. Swenk, *Catholic Spirit: Wesley, Whitefield, and the Quest for Evangelical Unity in Eighteenth-Century British Methodism* (Lanham, MD: Scarecrow, 2008). On Whitefield and Tennent, see Charles Maxson, *The Great Awakening in the Middle Colonies* (Chicago: University of Chicago Press, 1920). On styles of preaching, see Stephen Foster, *The Long Argument: English Puritanism and the Shaping of New England Culture, 1570–1700* (Chapel Hill: University of North Carolina Press, 1991); and Theodore Dwight Bozeman, *To Live Ancient Lives: The Primitivist Dimension in Puritanism* (Chapel Hill: University of North Carolina Press, 1988).

Whitefield's letters and sermons appeared during his lifetime in single publications from 1738 to 1770, over 275 prints (most from the years 1739 through 1744). Collections of these appeared immediately after his death. In addition to the journals and sermons Franklin published, I have used *Works of the Reverend George Whitefield* (6 vols.; London: Dilly, 1771–1772), *Fifteen Sermons Preached on Important Subjects by George Whitefield* (Philadelphia: Carey, 1794), the Gillies and Dallimore collections cited above, as well as the *Letters of George Whitefield, for the Period 1730–1742* (Edinburgh: Banner of Truth, 1976). Other letters have recently appeared but not yet been released, including a cache of letters that Bruce Hindmarsh is editing.

On print media in the Anglo-American Atlantic World, see Hugh Amory, "Reinventing the Colonial Book," in Hugh Amory, ed., *History of the Book in America, Volume One: The Colonial Book in the Atlantic World* (Chapel Hill: University of North Carolina Press, 2007), 26–54; T. H. Breen, *The Marketplace of Revolution: How Consumer Politics Shaped American Independence* (New York: Oxford University Press, 2004); E. Jennifer Monaghan, *Learning to Read and Write in Colonial America* (Amherst: University of Massachusetts Press, 2005); and James Raven, "The Importation of Books," in Amory, ed., *History of the Book*, 183–98.

Specific accounts of the meeting of the two men include Larry Gragg, "A Mere Civil Friendship: Franklin and Whitefield," *History Today* 28 (September 1, 1978): 574–79; Frank Lambert, "Subscribing for Profits and Piety: The Friendship of Benjamin Franklin and George Whitefield," *William and Mary Quarterly* 3rd ser. 50 (1993): 529–53; David T. Morgan, "A Most Unlikely Friendship: Benjamin Franklin and George Whitefield," *The Historian* 47 (1985): 208–18; William Pencak, "Beginning of a Beautiful Friendship: Benjamin Franklin, George Whitefield, The 'Dancing Blockheads,' and a Defense of the Meaner Sort," *Proteus* 19 (2002): 45–50; Samuel J. Rogal, "Toward a Mere Civil Friendship: Benjamin Franklin and George Whitefield," *Methodist History* 35 (1997): 233–43; and John R. Williams, "The Strange Case of Dr. Franklin and Mr. Whitefield," *Pennsylvania Magazine of History and Biography* 102 (1978): 399–421.

For the idea of the early modern man of letters as "self-fashioning," see Stephen Greenblatt, *Renaissance Self-Fashioning: From More to Shakespeare* (Chicago: University of Chicago Press, 1980); and Kenneth A. Lockridge, "Colonial Self-Fashioning: Paradoxes and Pathologies in the Construction of Genteel Identity in Eighteenth-Century America," in *Through A Glass Darkly: Reflections on Personal Identity in Early America*, ed. Ronald Hoffman, Mechal Sobel, and Fredrika J. Teute (Chapel Hill: University of North Carolina Press, 1997), 274–342. The term is now in wide use. See, e.g., Andrew W. Keitt, *Inventing the Sacred* (Leiden: Brill, 2005), 116–17.

For ideas of childhood, including the connection between religion and death, see Holly Brewer, *By Birth or Consent: Children, Law, and the Anglo-American Revolution in Authority* (Chapel Hill: University of North Carolina Press, 2005); Lawrence Stone, *The Family, Sex, and Marriage in England, 1500–1800* (London: Weidenfeld, 1977); and Edward Shorter, *The Making of the Modern Family* (New York: Basic Books, 1977).

Works on the Great Awakening always include some discussion of Whitefield. See, e.g., J. M. Bumsted, *The Great Awakening: The Beginnings of Evangelical Pietism in America* (Waltham, MA: Blaisdell, 1969); Richard L. Bushman, ed., *The Great Awakening: Documents on the Revival of Religion, 1740–1745* (New York: Atheneum, 1970); Edwin Gaustad, *The Great Awakening in New England* (New York: Harper, 1957); C. C. Goen, *Revivalism and Separatism in New England, 1740–1800: Strict Congregationalists and Separate Baptists in the Great Awakening* (Middleton, CT: Wesleyan University Press, 1987); Christopher Grasso, *A Speaking Aristocracy: Transforming Public Discourse in Eighteenth-Century Connecticut* (Chapel Hill: University of North Carolina Press, 1999); Alan Heimert and Perry Miller, eds., *The Great Awakening: Documents Illustrating the Crisis and Its Consequences* (Indianapolis: Bobbs-Merrill, 1967); Christine Heyrman, *Commerce and Culture: The Maritime Communities of Colonial Massachusetts, 1690–1750* (New York: Norton, 1984); Thomas S. Kidd, *The Great Awakening: The Roots of Evangelical Christianity in Colonial America* (New Haven: Yale University Press, 2007); Frank Lambert, *Inventing the "Great Awakening"* (Princeton: Princeton University Press, 1999); and Darrett Rutman, *The Great Awakening: Event and Exegesis* (New York: Wiley, 1970).

Both the Presbyterians and the Methodists claimed Whitefield as one of their own. See, e.g., Charles Wesley, *An Elegy on the Late Reverend George Whitefield, A.M., Who Died September 30, 1770, in the 56th Year of his Age* (Newark, NJ: Tuttle, 1809); and Robert Philip, *The Life and Times of the Reverend George Whitefield, M.A.* (New York: Appleton, 1838).

There is a separate literature on Franklin as a scientist. The two outstanding volumes devoted to his science are I. Bernard Cohen's collection of previously published articles, *Benjamin Franklin's Science* (Cambridge, MA: Harvard University Press, 1990), immensely detailed for so short and episodic a volume, and Joyce Chaplin, *The First Scientific American: Benjamin Franklin and the Pursuit of Genius* (Cambridge, MA: Harvard University Press, 2006), an essay on how Franklin became an icon of American ingenuity. A more general introduction is Thomas L. Hankins, *Science and the Enlightenment* (Cambridge: Cambridge University Press, 1985).

Just as he featured in just about every important colonial event in his time, so Franklin shows up in just about every colonial history. There is no room for them here, but the bibliographies and notes in the Franklin biographies cite them. Two are more recent than the biographies: Lorraine Smith Pangle, *The Political Philosophy of Benjamin Franklin* (Baltimore: Johns Hopkins University Press, 2007); and Alan Craig Houston, *Benjamin Franklin and the Politics of Improvement* (New Haven: Yale University Press, 2008).

INDEX